*Proceedings of the*

# Gökova Geometry-Topology Conference 2009

*Editors*

**Selman Akbulut**
Department of Mathematics, Michigan State University
East Lansing, MI 48824, USA

**Denis Auroux**
Department of Mathematics, Massachusetts Institute of Technology
Cambridge, MA 02139, USA

**Turgut Önder**
Department of Mathematics, Middle East Technical University
06531 Ankara, Turkey

GGT Gökova Geometry-Topology Conferences

International Press
www.intlpress.com

Published in 2010 by International Press, Somerville, Massachusetts, U.S.A. under license from Gökova Geometry-Topology Conferences.



ISBN 978-1-57146-199-5

Printed in the United States of America.

16$^{\text{th}}$ GÖKOVA GEOMETRY-TOPOLOGY CONFERENCE
May 25 – May 30, 2009

Participants of the 16$^{\text{th}}$ Gökova Geometry-Topology Conference

# PREFACE

This volume contains articles related to Sixteenth Gökova Conference which took place in May/2009 on the beautiful Gökova bay. These are stimulating survey articles as well as original research papers. We want to thank NSF and TÜBİTAK for funding this conference, we also thank International Press for printing and distributing these proceedings. A special thanks goes to Hotel Yücelen for being a great host.

March 2010 | The Editors

# *CONTENTS*

Proceedings of $16^{th}$ Gökova
Geometry-Topology Conference
pp. 1 – 21

# Group actions on 4-manifolds: some recent results and open questions

*Weimin Chen*

Abstract. A survey of finite group actions on symplectic 4-manifolds is given with a special emphasis on results and questions concerning smooth or symplectic classification of group actions, group actions and exotic smooth structures, and homological rigidity and boundedness of group actions. We also take this opportunity to include several results and questions which did not appear elsewhere.

Contents



## 1. Introduction

Let $X$ be a compact closed, oriented, simply connected, topological 4-manifold, and let $G$ be a finite group. Consider continuous actions of $G$ on $X$ which are orientation preserving, i.e., for every $g \in G$, the homeomorphism $g : X \to X$ is orientation preserving. We shall further assume that the actions of $G$ are locally linear, which means that for any $x \in X$ with nontrivial isotropy subgroup $G_x := \{g \in G | g \cdot x = x\} \neq \{1\}$, the action of $G_x$ near $x$ is modeled by a linear action, given by a faithful representation $\rho_x : G_x \to GL(4, \mathbb{R})$. Note that $\rho_x$ is uniquely determined up to linear conjugation because nonlinear similarity begins in dimension 6 (cf. [13]). For such an action of $G$ on $X$, one can associate to it the following data

$$\Gamma := (\rho, X^G, \{\rho_x\}),$$

where $\rho : G \to \text{Aut } (H^2(X), \cup)$ is the induced action on the second cohomology (note that it preserves the cup product $\cup$ on $H^2(X)$), $X^G := \{x \in X | G_x \neq \{1\}\}$ is the singular

*Key words and phrases.* Group actions; four-manifolds; symplectic.
The author was supported in part by NSF grant DMS-0603932.

set of the action, and $\rho_x : G_x \to GL(4,\mathbb{R})$ is the linear representation modeling the group action near $x$, $\forall x \in X^G$. The action is called *free* if $X^G = \emptyset$, *semifree* if $G_x = G$ for all $x \in X^G$, and *pseudofree* if $X^G$ is a finite set. A good starting point for understanding the action of $G$ on $X$ is the associated data $\Gamma := (\rho, X^G, \{\rho_x\})$.

The fundamental case to consider is when $G \equiv \mathbb{Z}_p$ is a cyclic group of prime order $p > 1$. In this case $X^G$ is simply the set of points fixed under $G$, which is a disjoint union of isolated points and 2-dimensional surfaces (because the action of $G$ preserves the orientation of $X$). For each $x \in X^G$, $G_x = G$ (i.e., the action is semifree), and moreover, if $x = m$ is an isolated fixed point, $\rho_x$ is given by an unordered pair of nonzero integers $(a_m, b_m)$ unique up to a simultaneous change of sign and congruence modulo $p$, and if $x \in Y$ lies in a fixed surface $Y$, $\rho_x$ is given by a nonzero integer $c_Y$, unique up to a change of sign and congruence modulo $p$. More precisely, in the former case $\rho_x : \lambda \mapsto diag(\lambda^{a_m}, \lambda^{b_m})$ and in the latter case $\rho_x : \lambda \mapsto diag(1, \lambda^{c_Y})$, where $\lambda$ is a $p$-th root of unity. Finally, for $\rho : G \to \text{Aut}\ (H^2(X), \cup)$, the corresponding integral $\mathbb{Z}_p$-representation on $H^2(X)$ can be decomposed into a direct sum

$$H^2(X) = \mathbb{Z}[\mathbb{Z}_p]^r \oplus \mathbb{Z}^t \oplus \mathbb{Z}[\mu_p]^s$$

for some integers $r, t, s \geq 0$, where the group ring $\mathbb{Z}[\mathbb{Z}_p]$ is the regular representation of $\mathbb{Z}$-rank $p$, $\mathbb{Z}$ is the trivial representation of $\mathbb{Z}$-rank 1, and $\mathbb{Z}[\mu_p]$ is the representation of cyclotomic type of $\mathbb{Z}$-rank $(p-1)$, which is the kernel of the augmentation homomorphism $\mathbb{Z}[\mathbb{Z}_p] \to \mathbb{Z}$ (see [41]). (Here $\mu_p \equiv \exp(2\pi i/p)$.)

We collect below the main constraints that have to be satisfied by $\Gamma := (\rho, X^G, \{\rho_x\})$ where $G \equiv \mathbb{Z}_p$. See [20], Section 3, for a more complete review.

**Proposition 1.1.** *(1)* $b_2(X) = rp + t + s(p-1)$.

*(2) (Lefschetz fixed point formula)* $\chi(X^G) = t - s + 2$.

*(3) (cf. [24]) If* $X^G \neq \emptyset$, *then* $b_1(X^G; \mathbb{Z}_p) = s$. *In particular, there are no summands of cyclotomic type in* $H^2(X)$ *(i.e.,* $s = 0$*) if the action is pseudofree with nonempty fixed point set.*

*(4) (cf. [24]) Suppose* $G = \langle g \rangle = \mathbb{Z}_2$. *If* $X^G$ *is empty or a finite set, then* $g^*\alpha \cup \alpha = 0 \pmod 2$ *for all* $\alpha \in H^2(X)$.

**Proposition 1.2.** *(1) (G-signature Theorem, [40, 35]) Assume each 2-dimensional component* $Y \subset X^G$ *is orientable. Set* $Sign(g, X) = tr(g)|_{H^{2,+}(X;\mathbb{R})} - tr(g)|_{H^{2,-}(X;\mathbb{R})}$, $\forall g \in G$. *Then*

$$Sign(g, X) = \sum_{m \in X^G} -\cot(\frac{a_m\pi}{p}) \cdot \cot(\frac{b_m\pi}{p}) + \sum_{Y \subset X^G} \csc^2(\frac{c_Y\pi}{p}) \cdot (Y \cdot Y),$$

*where* $Y \cdot Y$ *denotes the self-intersection number of* $Y$.

*(2) (Weaker version) Denote by Sign the signature of a space. Then*

$$p \cdot Sign(X/G) = Sign(X) + \sum_{m \in X^G} def_m + \sum_{Y \subset X^G} def_Y,$$

*where the terms* $def_m$ *and* $def_Y$ *(called* signature defect) *are given by the following formulae:*

$$def_m = \sum_{1 \neq \lambda \in \mathbb{C}, \lambda^p = 1} \frac{(1+\lambda^{a_m})(1+\lambda^{b_m})}{(1-\lambda^{a_m})(1-\lambda^{b_m})}$$

*and*

$$def_Y = \frac{p^2-1}{3} \cdot (Y \cdot Y).$$

It turns out that locally linear topological actions are largely determined by the associated data $\Gamma = (\rho, X^G, \{\rho_x\})$, at least for the case where $G \equiv \mathbb{Z}_p$ and the action is pseudofree (or free). First of all, there is the following realization result (for odd $p$) due to Edmonds and Ewing [27].

**Theorem 1.3.** *(Edmonds-Ewing [27]) Let $X$ be a closed, oriented, simply connected 4-manifold $X$, together with a representation of $\mathbb{Z}_p$ on $H^2(X)$ of the form $H^2(X) = \mathbb{Z}[\mathbb{Z}_p]^r \oplus \mathbb{Z}^t$, preserving the cup product, and with a candidate fixed point data for $t+2$ isolated fixed points satisfying the $G$-signature theorem and an additional torsion condition (which vanishes for relatively small $p$). Then there is a locally linear topological $\mathbb{Z}_p$-action on $X$ realizing the given data.*

The classification up to topological conjugacy was obtained by Wilczynski [64] and Bauer-Wilczynski [8], showing in particular that there are at most finitely many different conjugacy classes for a given data $\Gamma = (\rho, X^G, \{\rho_x\})$. Their work generalized earlier work of Hambleton-Kreck [37] on free actions.

Historically, the study of group actions has often focused on the investigation of certain manifolds which carry a natural geometric structure, where the geometric structure admits a large group of automorphisms. The most important example in this regard is the unit sphere of an Euclidean space, which inherits a linear structure from the ambient space. In the examples below, we shall examine some natural group actions in dimension 4.

**Example 1.1.** (1) Consider $\mathbb{S}^4$, the unit sphere in $\mathbb{R}^5$. Writing $\mathbb{R}^5 = \mathbb{C}^2 \times \mathbb{R}$, then every orientation-preserving linear $\mathbb{Z}_p$-action on $\mathbb{S}^4$ is conjugate to one of the following:

$$\lambda \cdot (z_1, z_2, x) \mapsto (\lambda z_1, \lambda^q z_2, x), \text{ where } \lambda^p = 1, 0 \leq q < p.$$

In the above model, the fixed point set consists of two isolated points, the north pole $(0,0,1)$ and the south pole $(0,0,-1)$, when $q \neq 0$, and is given by the embedded 2-sphere $z_1 = 0$ when $q = 0$. The local representations $\rho_x$ are given by $\lambda \cdot (z_1, z_2) \mapsto (\lambda z_1, \lambda^q z_2)$, $\lambda^p = 1$, and since $H^2(\mathbb{S}^4) = 0$, the induced representation $\rho$ on $H^2(\mathbb{S}^4)$ is trivial.

Though we are mainly concerned with orientation-preserving actions in this article, we point out that the only orientation-reversing linear action on $\mathbb{S}^4$ is the free involution given by the antipodal map.

(2) Consider the complex projective plane $\mathbb{CP}^2$. The models for complex linear $\mathbb{Z}_p$-actions are

$$\lambda \cdot [z_0 : z_1 : z_2] \mapsto [z_0 : \lambda^a z_1 : \lambda^b z_2], \text{ where } \lambda^p = 1, 0 \leq a < b < p.$$

These actions are all orientation-preserving and have a trivial representation on $H^2(\mathbb{CP}^2) = \mathbb{Z}$. The fixed point set consists of three isolated points $[0:0:1], [0:1:0]$, and $[1:0:0]$ when $a \neq 0$, with $\rho_x$ given by $(-b, a-b)$, $(-a, b-a)$ and $(a,b)$. When $a = 0$ (in which case one can assume $b = 1$), the fixed point set consists of one isolated point $[0:0:1]$ and an embedded 2-sphere, the complex line $z_2 = 0$. Note that when $p = 2$, $a = 0$ is the only possibility.

The anticomplex linear action is the involution $[z_0 : z_1 : z_2] \mapsto [\bar{z}_0 : \bar{z}_1 : \bar{z}_2]$, with the fixed point set being $\mathbb{RP}^2 = \{[x_0 : x_1 : x_2] | x_i \in \mathbb{R}\}$. The representation $\rho$ on $H^2(\mathbb{CP}^2)$ is given by multiplication by $-1$, and the action preserves the orientation.

**Question 1.1.** Is every $\mathbb{Z}_p$-action on $\mathbb{S}^4$ or $\mathbb{CP}^2$ conjugate to a linear action?

In the topological category the answer is yes for pseudofree actions; the case of $\mathbb{S}^4$ was due to Kwasik and Schultz [42], and the case of $\mathbb{CP}^2$ was due to Edmonds and Ewing [26] for the local representations (see also Hambleton-Lee [38]) and to Hambleton, Lee and Madsen [39] for the classification (see also Wilczynski [63]).

For non-pseudofree actions the answer is no. Giffen [32] gave the first examples of $\mathbb{Z}_p$-actions on $\mathbb{S}^4$ having a knotted $\mathbb{S}^2$ as the fixed point set. Hambleton, Lee and Madsen [39] used Giffen's examples to produce nonlinear $\mathbb{Z}_p \times \mathbb{Z}_p$-actions on $\mathbb{CP}^2$, but no examples of nonlinear $\mathbb{Z}_p$-actions on $\mathbb{CP}^2$ are known.

For free involutions on $\mathbb{S}^4$, there is a long story and the answer is no in both topological and smooth categories. Fintushel and Stern [29] gave the first example of an involution on $\mathbb{S}^4$ whose quotient is a fake smooth $\mathbb{RP}^4$. Later Gompf [33] showed that one of the earlier examples of fake smooth $\mathbb{RP}^4$'s due to Cappell and Shaneson [10] has universal covering diffeomorphic to $\mathbb{S}^4$; recently Akbulut [2] has shown that all of them have universal covering diffeomorphic to $\mathbb{S}^4$.

In Section 2, we discuss several results related to Question 1.1 on symplectic actions.

**Question 1.2.** Given a general simply connected 4-manifold $X$, how complex is the set of associated data $\Gamma = (\rho, X^G, \{\rho_x\})$ of all $\mathbb{Z}_p$-actions on $X$?

First, we note that for a 4-manifold $X$ which is not one of those with a relatively simple intersection form $(H^2(X), \cup)$ (e.g., $\mathbb{S}^4$, $\mathbb{CP}^2$, etc.), it is already a complicated algebraic problem to describe the representation $\rho$ induced by a $\mathbb{Z}_p$-action.

**Example 1.2.** A $K3$ surface is a simply connected complex surface with trivial canonical bundle. All $K3$ surfaces are Kähler, and have the same diffeomorphism type as smooth 4-manifolds [7]. An example of a $K3$ surface is given by the Fermat quartic $X_0 = \{z_0^4 + z_1^4 + z_2^4 + z_3^4 = 0\} \subset \mathbb{CP}^3$. The intersection form of a $K3$ surface is

$$H^2(X) = H \oplus H \oplus H \oplus E_8(-1) \oplus E_8(-1),$$

known as the $K3$ lattice.

Finite groups of automorphisms of a $K3$ surface have been extensively studied, particularly for the case of symplectic automorphisms, largely due to work of Nikulin [54] and Mukai [53] — these are the ones which act trivially on the canonical bundle. Nikulin classified the Abelian groups and showed that $\mathbb{Z}_p$ for $p = 2, 3, 5, 7$ are the only cyclic groups of a prime order. He also showed that the representation $\rho$ is uniquely determined up to conjugacy; this allows one to determine $\rho$ by examining certain specific examples, cf. [52, 31]. For $p = 2$, it is fairly easy to describe it — $\rho$ is the representation which fixes the 3 copies of $H$ and switches the 2 copies of $E_8(-1)$. For $p \geq 3$, it becomes fairly complicated to describe them. One often does it by describing the invariant lattice $H^2(X)^{\mathbb{Z}_p}$ and its orthogonal complement $\Omega_p := (H^2(X)^{\mathbb{Z}_p})^\perp$. For example, for $p = 3$, $H^2(X)^{\mathbb{Z}_p} = H \oplus H(3) \oplus H(3) \oplus A_2 \oplus A_2$ and $\Omega_p = K_{12}(-2)$, the Coxeter-Todd lattice of rank 12 with the bilinear form multiplied by $-2$.

Not much is known about the set of representations of $\mathbb{Z}_p$ on the $K3$ lattice which are induced by a pseudofree $\mathbb{Z}_p$-action. Besides the examples realized by a symplectic automorphism, a few more can be found in [20], Thm 1.8. Compare also Question 3.5.

Toward Question 1.2, one often focused on the following two specific questions: homological rigidity and boundedness of group actions. More precisely, homological rigidity concerns whether the representation $\rho$ is faithful, i.e., whether an action is trivial if the induced action on homology is trivial, while boundedness of actions asks whether a given manifold could admit group actions of arbitrarily large order. We should point out that these rigidity questions were also motivated by the corresponding results in dimension 2 (assuming the genus of the Riemann surface is at least two), both of which were due to Hurwitz.

In dimension 4, holomorphic actions tend to be rigid. $K3$ surfaces provide a classical example of homological rigidity; every homologically trivial automorphism of a $K3$ surface is trivial [7]. Peters [56] extended homological rigidity to elliptic surfaces (with a few exceptions), including in particular all simply connected elliptic surfaces. Concerning boundedness of actions, Hurwitz's theorem was extended to algebraic surfaces of general type, where the optimal bound was attained by Xiao [67, 68].

On the other hand, the following result of Edmonds [23] shows that topological actions are rather flexible and occur in abundance.

**Theorem 1.4.** *(Edmonds [23]) Let a closed, simply connected 4-manifold $X$ be given. For any prime number $p > 3$, there exists a locally linear, pseudofree and homologically trivial, topological $\mathbb{Z}_p$-action on $X$.*

However, for non-Abelian groups McCooey [50] established homological rigidity for locally linear topological actions.

In Section 4, we will discuss some results toward rigidity for symplectic actions.

Finally, we consider the following question which is only valid for smooth actions.

**Question 1.3.** Given a smooth 4-manifold, how much is the smooth structure reflected by its finite groups of (smooth) symmetries and vice versa?

In higher dimensions, this question was extensively studied in the case of homotopy spheres (cf. [58]). In Section 3, we will discuss some results in this direction for the case of homotopy $K3$ surfaces (smooth manifolds homeomorphic to a $K3$ surface).

The main technical tools for obtaining these results are Seiberg-Witten invariants and pseudoholomorphic curves, which are suitably adapted for the purpose of group actions. The readers are referred to the survey article [14] for an overview and the individual papers for more details.

## 2. Classificational results

Consider a smooth action of a finite group $G$ on $\mathbb{S}^4$ which has an isolated fixed point $x \in \mathbb{S}^4$. We are concerned with the question as whether the action is smoothly conjugate to a linear action. By taking out an invariant neighborhood of $x$, we obtain a smooth action of $G$ on the 4-ball $\mathbb{B}^4$ where the action is free and linear on the boundary. It is clear that the linearity of the action on $\mathbb{S}^4$ is equivalent to the linearity of the action on $\mathbb{B}^4$. Such a group $G$ which acts freely and linearly on $\mathbb{S}^3 = \partial\mathbb{B}^4$ has been classified, cf. e.g. [65]. In particular, its center is nontrivial, hence contains a cyclic subgroup of prime order. Applying the Smith theory [9] to this subgroup, it follows easily that the $G$-action on $\mathbb{B}^4$ is semifree and has a unique fixed point $y \in \mathbb{B}^4$. In the complement of an invariant neighborhood of $y$, the action is free with quotient an $h$-cobordism of $\mathbb{S}^3/G$, which, according to [43], is in fact an $s$-cobordism. Hence our problem is now reduced to whether a smooth $s$-cobordism of elliptic 3-manifolds $\mathbb{S}^3/G$ is necessarily trivial provided that its universal covering is trivial. There are nontrivial topological $s$-cobordisms of elliptic 3-manifolds [11, 42]. Some potentially nontrivial smooth examples were given by Cappell and Shaneson [12]; one of the Cappell-Shaneson examples in [12] was shown by Akbulut [1] to have a trivial universal covering.

In [15, 16] we showed that a symplectic $s$-cobordism of elliptic 3-manifolds (with standard structure near the boundary) is smoothly trivial. Using a theorem of Gromov [36] which says that a symplectic structure on $\mathbb{B}^4$ is standard if it is standard near the boundray, we can reformulate the results in [15, 16] as follows.

**Theorem 2.1.** *A smooth action of a finite group $G$ on $\mathbb{B}^4$, which is free and linear on the boundary and preserves the standard symplectic structure on $\mathbb{B}^4$, is conjugate to a linear action by a diffeomorphism equaling identity near the boundary.*

We conjectured in [16] that a smooth $s$-cobordism of elliptic 3-manifolds whose universal covering is trivial is symplectic, and suggested to attack this problem using near symplectic geometry [60]. One can reformulate the conjecture in [16] in terms of group actions on $\mathbb{S}^4$.

**Conjecture 2.2.** *A smooth finite group action on $\mathbb{S}^4$ with an isolated fixed point is smoothly conjugate to a linear action.*

Next, we give a corollary of Theorem 2.1, regarding group actions on $\mathbb{CP}^2$.

**Theorem 2.3.** *Suppose a finite group $G$ acts smoothly on $\mathbb{CP}^2$ which has an isolated fixed point and preserves a symplectic structure $\omega$. Then the $G$-action is smoothly conjugate to a linear action.*

*Proof.* Fix a $G$-equivariant, $\omega$-compatible almost complex structure $J$. The key point is to show that there is an embedded $J$-holomorphic 2-sphere $C$ representing a generator of $H_2(\mathbb{CP}^2)$, which is invariant under the action of $G$. Assume the existence of $C$ momentarily. Then the action of $G$ in an invariant neighborhood of $C$ is symplectomorphic to a linear action by the equivariant symplectic neighborhood theorem. Now by a theorem of Gromov [36], the complement of this neighborhood is symplectomorphic to $\mathbb{B}^4$ with the standard structure. By Theorem 2.1, the $G$-action is smoothly linear, which implies that the $G$-action on $\mathbb{CP}^2$ is smoothly linear.

It remains to show that such a $J$-holomorphic 2-sphere exists, for which we need to exploit the structure of $G$. Note that $G$ acts freely and linearly on $\mathbb{S}^3$ because the $G$-action on $\mathbb{CP}^2$ has an isolated fixed point. Such a group falls into two distinct classes: (i) cyclic groups, (ii) non-Abelian groups with a center containing a $\mathbb{Z}_2$-subgroup, cf. [65]. Considering case (i), let $g \in G$ be a generator. Then by the Lefschetz fixed point theorem, there must be at least two distinct points, say $x, y \in \mathbb{CP}^2$, that are fixed under $g$. By a theorem of Gromov [36], there is a unique embedded $J$-holomorphic 2-sphere $C$ passing through $x, y$ such that $C$ is a generator of $H_2(\mathbb{CP}^2)$. The uniqueness of $C$ implies that $C$ must be invariant under $g$, and this finishes the case (i). For case (ii), let $\tau \in G$ be the involution contained in the center of $G$. Then Proposition 1.1(4) implies that the fixed point set of $\tau$ contains a 2-dimensional component $C$, which must be a 2-sphere by Proposition 1.1(3). Furthermore, $C$ is clearly $J$-holomorphic. By Proposition 1.2(2), $C$ must be the only component in the fixed point set of $\tau$, and $C^2 = 1$. The latter implies that $C$ is a generator of $H_2(\mathbb{CP}^2)$. Finally, since $\tau$ is in the center of $G$, $C$ must be invariant under $G$. This finishes the case (ii). □

In view of Theorem 2.1, one naturally asks

**Question 2.1.** Let $\omega_0$ be the standard symplectic structure on $\mathbb{R}^4$. Suppose a finite group $G$ acts on $(\mathbb{B}^4, \omega_0)$ via symplectomorphisms, which are free and linear on the boundary of $\mathbb{B}^4$. Is the $G$-action conjugate to a linear action by a symplectomorphism equaling identity near the boundary?

We have a partial result, which was obtained in 2004 (cf. [18]).

**Theorem 2.4.** *(Chen [18]) Let $G$ be a cyclic or metacyclic finite group acting on $(\mathbb{B}^4, \omega_0)$ via symplectomorphisms which are linear near the boundary of $\mathbb{B}^4$. (We do not assume the action is free near the boundary.) Then the action of $G$ is conjugate to a linear action by a symplectomorphism of $(\mathbb{B}^4, \omega_0)$ which is identity near the boundary.*

*Proof.* The proof is an equivariant version of the proof in McDuff-Salamon [51] about the uniqueness of symplectic structures on the 4-ball.

First of all, we may identify $\mathbb{R}^4$ with $\mathbb{C}^2$ such that the action of $G$ near $\partial\mathbb{B}^4$ is complex linear.

Consider the linear action of $G$ on $\mathbb{C}^2$. We claim that there exists a decomposition $\mathbb{C}^2 = \mathbb{C} \oplus \mathbb{C}$ such that the action of $G$ on $\mathbb{C}^2$ can be extended to $\mathbb{S}^2 \times \mathbb{S}^2$, which is obtained via projectivizations of each factor in $\mathbb{C} \oplus \mathbb{C}$. The case when $G$ is cyclic is clear, we simply take $\mathbb{C}^2 = \mathbb{C} \oplus \mathbb{C}$ to be a decomposition into the eigenspaces of $G$. Now suppose $G$ is metacyclic, with a cyclic normal subgroup $H$ such that $G/H$ is cyclic. If each element of $H$ has two distinct eigenvalues, then we take $\mathbb{C}^2 = \mathbb{C} \oplus \mathbb{C}$ to be the decomposition into eigenspaces of $H$. Since $H$ is normal, any $g \in G$ either preserves $\mathbb{C} \oplus \mathbb{C}$ or switches the two factors. It is clear that the action of $G$ extends to an action on $\mathbb{S}^2 \times \mathbb{S}^2$. Finally, suppose the elements of $H$ have repeated eigenvalues. In this case $H$ lies in the center of $G$. We pick a $g \in G$ such that $[g] \in G/H$ generates $G/H$. Then a decomposition $\mathbb{C}^2 = \mathbb{C} \oplus \mathbb{C}$ into eigenspaces of $g$ will do.

With the preceding understood, we extend the $G$-action on $\mathbb{B}^4$ to the rest of $\mathbb{C}^2$ by linearity, and fix a decomposition $\mathbb{C}^2 = \mathbb{C} \oplus \mathbb{C}$ as described above. Then we compactify $\mathbb{C} \oplus \mathbb{C}$ via projectivizations of each factor into $M = \mathbb{S}^2 \times \mathbb{S}^2$. Note that $M$ has the standard split symplectic form $\omega_0$ which has equal areas on $\mathbb{S}^2 \times \{\infty\}$ and $\{\infty\} \times \mathbb{S}^2$. Moreover, the action of $G$ extends naturally to an action on $(M, \omega_0)$ which is linear near $\mathbb{S}^2 \times \{\infty\} \cup \{\infty\} \times \mathbb{S}^2$.

Fix an $\omega_0$-compatible almost complex structure $J$ on $M$ such that $J$ is $G$-equivariant and is the product complex structure near $\mathbb{S}^2 \times \{\infty\} \cup \{\infty\} \times \mathbb{S}^2$. Then there are two families of $J$-holomorphic embedded 2-spheres representing the classes of $\mathbb{S}^2 \times \{\infty\}$ and $\{\infty\} \times \mathbb{S}^2$ respectively. Call an element in the former an $A$-curve and an element in the latter a $B$-curve. Note that under the action of $G$, an $A$-curve is either sent to an $A$-curve or a $B$-curve.

Now consider $\mathbb{S}^2 \times \mathbb{S}^2$, which is equipped with the corresponding linear $G$-action, and is given with a compatible identification of $\mathbb{S}^2 \times \{\infty\}$ and $\{\infty\} \times \mathbb{S}^2$ to those in $M$. Then there is a diffeomorphism $\psi : \mathbb{S}^2 \times \mathbb{S}^2 \rightarrow M$ defined by $\psi(z, w) = \text{image } u_w \cap \text{image } v_z$, where $u_w$ is the $A$-curve passing through $\{\infty\} \times \{w\} \in M$ and $v_z$ is the $B$-curve passing through $\{z\} \times \{\infty\} \in M$. By way of construction, $\psi$ is identity along the union of $\mathbb{S}^2 \times \{\infty\}$ and $\{\infty\} \times \mathbb{S}^2$. Moreover, note that for any $g \in G$, either $g(u_w) = u_{gw}$ and $g(v_z) = v_{gz}$, or $g(u_w) = v_{gw}$ and $g(v_z) = u_{gz}$, depending on whether $g$ preserves $\mathbb{S}^2 \times \{\infty\}$ and $\{\infty\} \times \mathbb{S}^2$ or switches them. In any event, it follows easily that $\psi$ is equivariant with respect to the linear $G$-action on $\mathbb{S}^2 \times \mathbb{S}^2$ and the $G$-action on $M$.

The diffeomorphism $\psi$ is further modified into a diffeomorphism $\psi' : \mathbb{S}^2 \times \mathbb{S}^2 \rightarrow M$ which is identity near $\mathbb{S}^2 \times \{\infty\} \cup \{\infty\} \times \mathbb{S}^2$ and equals $\psi$ outside of a neighborhood of the union of $\mathbb{S}^2 \times \{\infty\}$ and $\{\infty\} \times \mathbb{S}^2$. To define $\psi'$, note that in a neighborhood of $\mathbb{S}^2 \times \{\infty\}$, $\psi$ is given by $(z, w) \mapsto (\phi(z, w), w)$ for a family of holomorphic functions $w \mapsto \phi(z, w)$, $z \in \mathbb{S}^2$, with $\phi(z, \infty) = z$, and similarly in a neighborhood of $\{\infty\} \times \mathbb{S}^2$, $\psi$ is given by $(z, w) \mapsto (z, \phi'(z, w))$ for a family of holomorphic functions $z \mapsto \phi'(z, w)$, $w \in \mathbb{S}^2$, with $\phi'(\infty, w) = w$. We fix a cutoff function $\beta$ which equals $\infty$ near $\infty$ and equals 1 outside a neighborhood of $\infty$, and define $\rho(x) = \beta(|x|)x$ for $x$ in a neighborhood of $\infty \in \mathbb{S}^2$. With these understood, $\psi'$ is defined by $(z, w) \mapsto (\phi(z, \rho(w)), w)$ in a neighborhood of $\mathbb{S}^2 \times \{\infty\}$ and by $(z, w) \mapsto (z, \phi'(\rho(z), w))$ in a neighborhood of $\{\infty\} \times \mathbb{S}^2$. We claim that $\psi'$ is also $G$-equivariant. To see this, note that the linear $G$-action on $\mathbb{S}^2 \times \mathbb{S}^2$ is given by either $g \cdot (z, w) = (az, bw)$ or $g \cdot (z, w) = (aw, bz)$ for some $a, b \in \mathbb{C}$ such that $|a| = |b| = 1$. The fact that $\psi$ is equivariant and that the action of $G$ on $M$ is linear

near $\mathbb{S}^2 \times \{\infty\} \cup \{\infty\} \times \mathbb{S}^2 \subset M$ implies that $\phi(az, bw) = a\phi(z, w)$ in the former case and $\phi'(aw, bz) = b\phi(z, w)$ in the latter case. It is easy to check that $\phi(az, \rho(bw)) = a\phi(z, \rho(w))$ in the former case and $\phi'(\rho(aw), bz) = b\phi(z, \rho(w))$ in the latter case, which implies that $\psi'$ is $G$-equivariant.

Now apply Moser's argument to the following family of symplectic forms on $\mathbb{S}^2 \times \mathbb{S}^2$

$$\omega_t = (1-t)\omega_0 + t(\psi')^*\omega_0,\ 0 \leq t \leq 1.$$

Note that each $\omega_t$ is $G$-equivariant with respect to the linear $G$-action on $\mathbb{S}^2 \times \mathbb{S}^2$, and $\omega_t = \omega_0$ for all $t \in [0,1]$ near $\mathbb{S}^2 \times \{\infty\} \cup \{\infty\} \times \mathbb{S}^2$. It follows easily that there exists a 1-form $\alpha$ which obeys: (1) $\frac{d}{dt}\omega_t = d\alpha$, (2) $\alpha$ vanishes near $\mathbb{S}^2 \times \{\infty\} \cup \{\infty\} \times \mathbb{S}^2$, and (3) $\alpha$ is $G$-equivariant. Let $X_t$ be the time-dependent vectorfield on $\mathbb{S}^2 \times \mathbb{S}^2$ which is defined by the equation $\alpha + i(X_t)\omega_t = 0$, then $X_t$ is clearly also $G$-equivariant, and vanishes near $\mathbb{S}^2 \times \{\infty\} \cup \{\infty\} \times \mathbb{S}^2$. Let $\psi_1$ be the time-one map generated by $X_t$, then $\psi_1$ is identity near $\mathbb{S}^2 \times \{\infty\} \cup \{\infty\} \times \mathbb{S}^2$ and is $G$-equivariant with respect to the linear $G$-action on $\mathbb{S}^2 \times \mathbb{S}^2$. Moreover, $\psi_1^*((\psi')^*\omega_0) = \omega_0$.

Define $\Psi : \mathbb{C}^2 \to \mathbb{C}^2$ to be the restriction of $\psi' \circ \psi_1$ to $\mathbb{C} \oplus \mathbb{C} \subset \mathbb{S}^2 \times \mathbb{S}^2$ which is identity near infinity. Then $\Psi^*\omega_0 = \omega_0$ and $\Psi$ is $G$-equivariant where $G$ acts on the domain $\mathbb{C}^2$ linearly and acts on the range $\mathbb{C}^2$ by the natural extension of the action of $G$ on $\mathbb{B}^4$. Now apply the following equivariant version of Lemma 9.4.10 in McDuff-Salamon [51] to $f \equiv \Psi^{-1}|_{\mathbb{C}^2 \setminus V}$, where $V \subset \mathbb{B}^4$ is a compact, $G$-invariant convex sub-domain such that $0 \in int(V)$ and $G$ acts linearly on $\mathbb{B}^4 \setminus V$:

*Let $G$ be a compact Lie group acting on $(\mathbb{R}^{2n}, \omega_0)$ via linear symplectomorphisms. Suppose that $V \subset \mathbb{R}^{2n}$ is a $G$-invariant, star-shaped compact set with $0 \in int(V)$ and $f : \mathbb{R}^{2n} \setminus V \to \mathbb{R}^{2n}$ is a $G$-equivariant symplectic embedding equaling identity near infinity. Then, for every $G$-invariant open neighborhood $W \subset \mathbb{R}^{2n}$ of $V$, there exists a $G$-equivariant symplectomorphism $g : \mathbb{R}^{2n} \to \mathbb{R}^{2n}$ such that $g|_{\mathbb{R}^{2n} \setminus W} = f$.*

Let $g : \mathbb{C}^2 \to \mathbb{C}^2$ be the $G$-equivariant symplectomorphism obtained from the lemma. Then $\Phi \equiv (\Psi \circ g)|_{\mathbb{B}^4} : \mathbb{B}^4 \to \mathbb{B}^4$ is identity near the boundary and satisfies $\Phi^*\omega_0 = \omega_0$, and is $G$-equivariant where the action of $G$ on the domain of $\Phi$ is linear while on the range of $\Phi$ it is the given action on $\mathbb{B}^4$. □

There are two immediate corollaries. As argued in Theorem 2.3, we obtain a classification for certain finite subgroups of the symplectomorphism group of $\mathbb{CP}^2$.

**Theorem 2.5.** *(Chen [18]) Let $G \subset Symp(\mathbb{CP}^2, \omega_0)$ be a finite subgroup which is either cyclic or metacyclic with a nonempty fixed point set. Then $G$ is conjugate in $Symp(\mathbb{CP}^2, \omega_0)$ to a subgroup of $PU(3)$. Here $\omega_0$ is the standard Kähler structure.*

With Example 3.5 in [14], we obtain a classification of symplectic structures on weighted $\mathbb{CP}^2$'s, generalizing the corresponding theorem for $\mathbb{CP}^2$ due to Gromov and Taubes [36, 59].

**Theorem 2.6.** *(Chen [18]) Two symplectic forms $\omega_1, \omega_2$ on a weighted projective plane $\mathbb{P}(d_1, d_2, d_3)$ are cohomologous if and only if there exists a homologically trivial self-diffeomorphism $\psi$ of $\mathbb{P}(d_1, d_2, d_3)$ such that $\psi^*\omega_2 = \omega_1$.*

Here $\mathbb{P}(d_1, d_2, d_3)$ is the orbifold $\mathbb{S}^5/\mathbb{S}^1$, where $\mathbb{S}^1$ acts linearly with weights $d_1, d_2, d_3$, i.e., $\lambda \cdot (z_1, z_2, z_3) = (\lambda^{d_1} z_1, \lambda^{d_2} z_2, \lambda^{d_3} z_3)$. (We assume $d_1, d_2, d_3$ are relatively prime. )

Note that in all of the results discussed above concerning group actions on $\mathbb{CP}^2$, there is always the assumption that the action has a fixed point. The only linear action on $\mathbb{CP}^2$ which is pseudofree without a fixed point is by the metacyclic group $\Gamma_{n,r}$, where

$$\Gamma_{n,r} = \{x, u \mid x^n = u^3 = 1, u^{-1}xu = x^r, r^2 + r + 1 \equiv 0 \pmod{n}\}.$$

A linear model for the actions of $\Gamma_{n,r}$ on $\mathbb{CP}^2$ is given below where $\lambda^n = 1$:

$$x \cdot [z_0 : z_1 : z_2] = [z_0 : \lambda^{-r} z_1 : \lambda z_2] \text{ and } u \cdot [z_0 : z_1 : z_2] = [z_2 : z_0 : z_1].$$

It was shown in Wilczynski [63] that any pseudofree, locally linear topological action of $\Gamma_{n,r}$ on $\mathbb{CP}^2$ is conjugate to a linear action.

**Question 2.2.** (Chen [18]) Is every pseudofree symplectic $\Gamma_{n,r}$-action on $\mathbb{CP}^2$ smoothly conjugate to a linear action?

## 3. Symmetries and exotic smooth structures

A classical theorem in Riemannian geometry says that if a compact Lie group $G$ acts smoothly and effectively on a compact closed $n$-dimensional manifold $M^n$, then the dimension of $G$ can not exceed $n(n+1)/2$, where if indeed $\dim G = n(n+1)/2$, $M^n$ must be diffeomorphic to $\mathbb{S}^n$ or $\mathbb{RP}^n$. In dimensions greater than 6, there exist exotic spheres, i.e., smooth manifolds which are homeomorphic but not diffeomorphic to the standard sphere $\mathbb{S}^n$. Thus the said theorem in Riemannian geometry gives us a criterion for distinguishing the standard sphere from an exotic one in terms of groups of symmetries — the standard sphere has the largest group of symmetries. On the other hand, there are exotic spheres whose symmetry groups are found to be very restricted. For example, Lawson and Yau [44] showed that there are exotic spheres which do not even support actions of small groups such as $\mathbb{S}^3$ or $SO(3)$.

In Chen-Kwasik [20, 21], we investigated these phenomena in dimension 4, using homotopy $K3$ surfaces as the testing ground. On the one hand, the finite symplectic automorphism groups of a $K3$ surface (called a *symplectic* $K3$ *group*) have been classified, cf. e.g. Mukai [53], which provided us with a large list of finite groups that can act at least continuously on a homotopy $K3$ surface. On the other hand, there are well-established methods, cf. e.g. [30, 59], to construct and analyze exotic smooth structures on a homotopy $K3$ surface.

The basic idea of our approach in [20, 21] can be summarized as follows. Let $X$ be a homotopy $K3$ surface which admits a symplectic structure. Then deep work of Taubes [59] implies that the set of Seiberg-Witten basic classes of $X$ spans an isotropic sublattice $L_X$ of $H^2(X)$ (with respect to the cup product), so that its rank, denoted by $r_X$, must range from 0 to 3, where $3 = \min(b_2^+, b_2^-)$. The rank $r_X$ of the lattice $L_X$ gives a rough measurement of the exoticness of the smooth structure of $X$, with $r_X = 0$ being the least exotic and with $r_X = 3$ being the most exotic. Furthermore, $r_X = 0$ implies that $X$ has a trivial canonical bundle, and the standard $K3$ is the only known example to have $r_X = 0$. (We will call $X$ "standard" if $r_X = 0$.)

The link between the exoticness $r_X$ and symmetry groups of $X$ is made as follows. Suppose a finite group $G$ acts smoothly on $X$. Then the lattice $L_X$ is invariant under the $G$-action, and since $L_X$ is isotropic, $L_X \subset L_X^\perp$, so that there is an induced $G$-action on $L_X^\perp/L_X$. In other words, the $G$-representation $\rho$ on $H^2(X)$ breaks up into two smaller pieces, one on $L_X$ and the other on $L_X^\perp/L_X$. If one further assumes that $G$ preserves a symplectic structure on $X$, then the pseudoholomorphic curve techniques developed in [19] allow one to analyze the fixed point set structure $(X^G, \{\rho_x\})$, and then through the Lefschetz fixed point theorem and the $G$-signature theorem, one obtains information about the $G$-representation $\rho$ on $H^2(X)$.

With the preceding understood, we investigated the following question in [20]. Suppose we change the smooth structure of the standard $K3$. Is it possible to arrange it so that some of the group actions on the standard $K3$ are no longer smoothable, and/or some of the groups which can act smoothly on the standard $K3$ can no longer act smoothly on the exotic $K3$?

**Theorem 3.1.** *(Chen-Kwasik [20]) There exist infinitely many symplectic homotopy $K3$ surfaces $X_\alpha$ with the exoticness $r_{X_\alpha} = 3$, such that the following statements hold.*

(1) *Let $g$ be a holomorphic automorphism of prime order $\geq 7$ of a $K3$ surface, or an anti-holomorphic involution of a $K3$ surface whose fixed-point set contains a component of genus $\geq 2$. Then $g$ is not smoothable on $X_\alpha$.*

(2) *Let $G$ be a finite group whose commutator $[G, G]$ contains a subgroup isomorphic to $(\mathbb{Z}_2)^4$ or $Q_8 = \{i, j, k | i^2 = j^2 = k^2 = -1, ij = k, jk = i, ki = j\}$, where in the case of $Q_8$ the elements of order 4 in the subgroup are conjugate in $G$. Then there are no effective smooth $G$-actions on $X_\alpha$.*

**Remark 3.1.** (1) The automorphism $g$ in Theorem 3.1(1) does exist [54, 55, 48], which provides the first examples of relatively non-smoothable, locally linear topological actions on a 4-manifold, i.e., examples of actions which are smoothable with respect to one smooth structure but non-smoothable with respect to some other smooth structures. There were previously known examples of absolutely non-smoothable actions, for example, there exists a locally linear topological action of order 5 on $\mathbb{CP}^2 \# \mathbb{CP}^2$ which is not smoothable with respect to any smooth structure on $\mathbb{CP}^2 \# \mathbb{CP}^2$, see [27, 38]. For more recent examples of absolutely non-smoothable actions, see e.g. [46, 47]. See also the recent survey by Edmonds [25].

(2) Among the 11 maximal symplectic $K3$ groups (see [53, 66] for a complete list), there are 6 of them satisfying the assumption in Theorem 3.1(2). They are

$$M_{20}, F_{384}, A_{4,4}, T_{192}, H_{192}, T_{48}.$$

On the other hand, the symplectic $K3$ group of smallest order which satisfies the assumption in Theorem 3.1(2) is the binary tetrahedral group $T_{24}$ of order 24.

(3) The $X_\alpha$'s in Theorem 3.1 were originally due to Fintushel and Stern [30]. The crucial fact we established in [20] (see Lemma 4.2 therein) is that

$$L_{X_\alpha}^{\perp}/L_{X_\alpha} = E_8(-1) \oplus E_8(-1).$$

This implies that any smooth $G$-action on $X_\alpha$ induces a $G$-representation on the $E_8$-lattice, which yields strong informations about the group $G$.

We also analyzed in [20] the fixed point set $X^G$ and local representations $\{\rho_x\}$ of the symplectic $\mathbb{Z}_p$-actions on $X_\alpha$ which induce a nontrivial action on each of the two $E_8$-lattices in $L_{X_\alpha}^{\perp}/L_{X_\alpha} = E_8(-1) \oplus E_8(-1)$, cf. Theorem 1.8 therein. The following question remains open.

**Question 3.1.** (Chen-Kwasik [20]) Does there exist an exotic $K3$ surface which admits no symplectic $\mathbb{Z}_p$-actions for some $p \leq 19$?

We remark that for every prime number $p \leq 19$, there is a holomorphic $\mathbb{Z}_p$-action on a $K3$ surface [54, 48].

Now we review the main results in [21]. Suppose we are given a symplectic homotopy $K3$ surface $X$ which admits a symplectic action of a "large" $K3$ group (e.g., one of the 11 maximal symplectic $K3$ groups). What can be said about the smooth structure of $X$ (e.g., in terms of the exoticness $r_X$)?

**Theorem 3.2.** *(Chen-Kwasik [21]) Let G be one of the following maximal symplectic $K3$ groups:*

$$L_2(7), A_6, M_{20}, A_{4,4}, T_{192}, T_{48}$$

*and let $X$ be a symplectic homotopy $K3$ surface. If $X$ admits an effective symplectic $G$-action, then $X$ must be "standard" (i.e., $r_X = 0$).*

The basic idea of the proof may be summarized as follows. Using the techniques developed in [19] and exploiting various features of the structure of $G$, one first determines the possible fixed point set of an arbitrary element $g \in G$, from which the trace $tr(g)$ of $g$ on $H^*(X;\mathbb{R})$ can be computed using the Lefschetz fixed point theorem. This leads to an estimate on

$$\dim(H^*(X;\mathbb{R}))^G = \frac{1}{|G|}\sum_{g\in G} tr(g).$$

On the other hand, the following basic inequality

$$\dim(L_X \otimes_{\mathbb{Z}} \mathbb{R})^G \leq \min(b_2^+(X/G), b_2^-(X/G))$$

plus the identity $\dim(H^*(X;\mathbb{R}))^G = 2 + b_2^+(X/G) + b_2^-(X/G)$ allows one to obtain information about $\dim(L_X \otimes_{\mathbb{Z}} \mathbb{R})^G$ and $r_X = \text{rank } L_X$.

For an illustration we consider the case where $G$ is a non-Abelian simple group. It is easily seen that in this case $b_2^+(X/G) = 3$ and $r_X = \dim(L_X \otimes_{\mathbb{Z}} \mathbb{R})^G$. The above basic inequality then becomes

$$r_X \leq \min(3, \dim(H^*(X;\mathbb{R}))^G - 5).$$

There are three non-Abelian simple $K3$ groups: $L_2(7), A_5$ and $A_6$. For the case where $G = L_2(7)$ or $A_6$, one can show that $\dim(H^*(X;\mathbb{R}))^G = 5$, so that $r_X = 0$. For $G = A_5$, one can only show that $\dim(H^*(X;\mathbb{R}))^G \leq 8$, so that the basic inequality gives only $r_X \leq 3$, which does not yield any restriction on the exoticness $r_X$.

**Question 3.2.** (Chen-Kwasik [21]) Suppose a symplectic homotopy $K3$ surface $X$ admits a symplectic $A_5$-action. Is $X$ necessarily "standard" (i.e., $r_X = 0$)?

Theorem 3.2 gives a characterization of "standard" symplectic homotopy $K3$ surfaces in terms of finite symplectic symmetry groups. It naturally raises the question as what can be said about the finite symplectic symmetry groups of a "standard" symplectic homotopy $K3$ surface. The answer is given in the following theorem.

**Theorem 3.3.** *(Chen-Kwasik [21]) Let $X$ be a "standard" symplectic homotopy $K3$ surface (i.e. $r_X = 0$) and let $G$ be a finite group acting on $X$ via symplectic symmetries. Then there exists a short exact sequence of finite groups*

$$1 \rightarrow G_0 \rightarrow G \rightarrow G^0 \rightarrow 1,$$

*where $G^0$ is cyclic and $G_0$ is a symplectic $K3$ group, such that $G_0$ is characterized as the maximal subgroup of $G$ with the property $b_2^+(X/G_0) = 3$. Moreover, the action of $G_0$ on $X$ has the same fixed-point set structure $(X^G, \{\rho_x\})$ of a symplectic holomorphic action of $G_0$ on a $K3$ surface.*

Theorem 3.3 raises the following questions.

**Question 3.3.** (Chen-Kwasik [21]) Are there any finite groups other than a $K3$ group which can act smoothly or symplectically on a homotopy $K3$ surface?

If a finite group $G$ other than a $K3$ group acts symplectically on a homotopy $K3$ surface $X$ such that $b_2^+(X/G) = 3$, then $X$ must be an exotic $K3$. It would be interesting to see a construction of such a homotopy $K3$ surface. Similarly,

**Question 3.4.** (Chen-Kwasik [21]) Are there any smooth or symplectic actions of a $K3$ group on a homotopy $K3$ surface which have a fixed-point set structure $(X^G, \{\rho_x\})$ different than a holomorphic action?

According to Theorem 3.3, a symplectic $\mathbb{Z}_p$-action on a "standard" symplectic homotopy $K3$ surface $X$ with $b_2^+(X/G) = 3$ should have the same fixed-point set structure $(X^G, \{\rho_x\})$ of a symplectic automorphism of order $p$ of a $K3$ surface. A natural question is: how much will this determine the induced representation $\rho$ on the $K3$ lattice? In particular, does it determine $\rho$ completely? Note that in the holomorphic case, the affirmative answer of Nikulin in [54] relied heavily on the global Torelli theorem for $K3$ surfaces. (Note that over $\mathbb{Q}$, the induced action is completely determined.)

**Question 3.5.** Let a $\mathbb{Z}_p$-action on a homotopy $K3$-surface (locally linear topological, smooth, or symplectic) be given which has the same fixed-point set structure $(X^G, \{\rho_x\})$ of an order $p$ symplectic automorphism of a $K3$ surface. What can be said about the induced representation $\rho$ on the $K3$ lattice?

For $p = 2$, we actually have an affirmative answer (however, for homeomorphism types up to conjugacy the answer seems not that simple, cf. [22]).

**Proposition 3.4.** *Let $g$ be a locally linear, orientation-preserving topological involution on a homotopy $K3$ surface $X$ which has 8 isolated fixed points. Then the induced action of $g$ on the $K3$ lattice is the same as that of a symplectic automorphism of order 2 of a $K3$ surface.*

*Proof.* First of all, the Lefschetz fixed point theorem and the $G$-signature theorem imply that $\text{Sign}(X/\langle g\rangle) = -8$ and $b_2^+(X/\langle g\rangle) = 3$. With this understood, the proof is based on the following theorem of Nikulin [54].

Denote by $L$ the $K3$ lattice. Suppose $G \subset O(L)$ is a finite subgroup of isometries of $L$. Let $L^G$ denote the sublattice of $L$ fixed by $G$, and let $S_G \equiv (L^G)^\perp$ be the orthogonal complement of $L^G$ in $L$.

**Theorem 3.5.** *(Nikulin, Theorem 4.3 in [54]) A finite subgroup $G \subset O(L)$ is realized as the induced action of a finite symplectic automorphism group of a $K3$ surface if the following conditions hold:*

(a) *$S_G$ is negative definite,*
(b) *$S_G$ does not have any elements with square $-2$,*
(c) *rank $S_G \leq 18$.*

Now let $G \equiv \mathbb{Z}_2 \subset O(L)$ be the subgroup of isometries of the $K3$ lattice induced by $g$. We shall verify that the conditions (a), (b), (c) in Theorem 3.5 are satisfied. First of all, $S_G$ is negative definite because $b_2^+(X/\langle g\rangle) = 3 = b_2^+(X)$. As for condition (c),

$$\text{rank } S_G = 22 - 2b_2^+(X/\langle g\rangle) + \text{Sign}(X/\langle g\rangle) = 8 \leq 18.$$

The verification of condition (b) is more involved. First of all, recall that by a theorem of Kwasik and Schultz (cf. [41]), the integral representation of $G \equiv \mathbb{Z}_2$ on $L$ can be expressed as a sum of copies of the group ring $\mathbb{Z}[\mathbb{Z}_2]$, the trivial representation $\mathbb{Z}$, and the representation $\mathbb{Z}[\mu_2]$ of cyclotomic type. By Proposition 1.1(3), there should be no summands of cyclotomic type in the representation. With this said, there are elements $x_1, \cdots, x_n, y_1, \cdots, y_m$ of $L$ such that the $x_i$'s are fixed by $g$ and $x_1, \cdots, x_n, y_1, \cdots, y_m, gy_1, \cdots, gy_m$ form a $\mathbb{Z}$-basis of $L$. Now let $u \in S_G$ be any element. Note that $u$ obeys $gu = -u$. We write

$$u = a_1x_1 + \cdots + a_nx_n + b_1y_1 + \cdots + b_my_m + c_1gy_1 + \cdots + c_mgy_m.$$

Then $gu = -u$ is equivalent to

$$2a_1x_1+\cdots+2a_nx_n+(b_1+c_1)y_1+\cdots+(b_m+c_m)y_m+(b_1+c_1)gy_1+\cdots+(b_m+c_m)gy_m = 0,$$

which implies that $a_i = 0$ for all $i$, and $b_j = -c_j$ for all $j$. Consequently,

$$u = v - gv, \text{ where } v = b_1y_1 + \cdots + b_my_m.$$

Computing the square of $u$, we have

$$(u, u) = (v, v) - (v, gv) - (gv, v) + (gv, gv) = 2(v, v) - 2(v, gv).$$

By Proposition 1.1(4), $(v, gv) = 0 \pmod 2$. On the other hand, $(v, v) = 0 \pmod 2$ since $L$ is even. This shows that $(u, u) = 0 \pmod 4$. Particularly, $u$ does not have square $-2$, and the condition (b) in Theorem 3.5 is verified. □

One way to construct symplectic $\mathbb{Z}_p$-actions on the standard $K3$ surface is to fix a $C^\infty$-elliptic fibration on the $K3$ surface and consider fibration-preserving $\mathbb{Z}_p$-actions. (It is a standard argument to show that these actions are symplectic.)

**Question 3.6.** Suppose a smooth $\mathbb{Z}_p$-action on the standard $K3$ surface preserves a $C^\infty$-elliptic fibration. Is it necessarily smoothly conjugate to a holomorphic action? What can be said about the induced action on the $K3$ lattice?

Finally, call a finite group $G$ (particularly a $K3$ group) "small" if $G$ can act on a symplectic homotopy $K3$ surface $X$ with $r_X = 3$ via symplectic symmetries.

**Question 3.7.** (Chen-Kwasik [21]) What can be said about the set of "small" groups?

In [21] we showed that $(\mathbb{Z}_2)^3$ is "small".

**Theorem 3.6.** *(Chen-Kwasik [21]) Let $G \equiv (\mathbb{Z}_2)^3$. There exists an infinite family of distinct symplectic homotopy $K3$ surfaces with maximal exoticness (i.e. $r_X = 3$), such that each member of the exotic $K3$'s admits an effective $G$-action via symplectic symmetries. Moreover, the $G$-action is pseudofree and induces a trivial action on the lattice $L_X$ of the Seiberg-Witten basic classes.*

## 4. Rigidity of group actions

The main theme in [17] concerns boundedness of $\mathbb{Z}_p$-actions on a given 4-manifold. Of course, there are necessary conditions that have to be imposed. By Edmonds' theorem (cf. Theorem 1.4), it is necessary to consider at least smooth actions. On the other hand, it is clear that if the manifold admits a smooth $\mathbb{S}^1$-action, there is no chance for a bound on the order to exist. With this understood, the main question in [17] can be formulated as follows.

**Question 4.1.** (Chen [17]) Let $X$ be a smooth 4-manifold which admits no smooth $\mathbb{S}^1$-actions. Does there exist a constant $C > 0$ such that there are no nontrivial smooth $\mathbb{Z}_p$-actions on $X$ for any prime number $p > C$? Moreover, suppose such a constant $C$ does exist, what structures of $X$ does $C$ depend on?

Before we state the theorems, it is helpful to recall the relevant results regarding the existence of smooth $\mathbb{S}^1$-actions on 4-manifolds. First, with the resolution of the 3-dimensional Poincaré conjecture by Perelman, an old theorem of Fintushel [28] may be strengthened to the following: a simply connected 4-manifold admitting a smooth $\mathbb{S}^1$-action must be a connected sum of copies of $\mathbb{S}^4$, $\pm\mathbb{CP}^2$, and $\mathbb{S}^2 \times \mathbb{S}^2$. For the non-simply connected case, one has the following useful criterion due to Baldridge [4]: Let $X$ be a 4-manifold admitting a smooth $\mathbb{S}^1$-action with nonempty fixed point set. Then $X$ has vanishing Seiberg-Witten invariant when $b_2^+ > 1$, and when $b_2^+ = 1$ and

$X$ is symplectic, $X$ is diffeomorphic to a rational or ruled surface. For fixed-point free smooth $\mathbb{S}^1$-actions, formulas relating the Seiberg-Witten invariants of the 4-manifold and the quotient 3-orbifold were given in Baldridge [5, 6]. Note that if a 4-manifold admits a fixed-point free $\mathbb{S}^1$-action, the Euler characteristic and the signature of the manifold must vanish.

With this understood, it is instructive to first look at the case of holomorphic actions, as these are the primary source of smooth actions on 4-manifolds. Based on Xiao's generalization of Hurwitz's theorem in [67, 68] and the techniques developed by Ueno [61] and Peters [56] regarding homological rigidity of holomorphic actions on elliptic surfaces, we obtained the following result.

**Theorem 4.1.** *(Chen [17]) Let $X$ be a compact complex surface with Kodaira dimension $\kappa(X) \geq 0$. Suppose $X$ does not admit any holomorphic $\mathbb{S}^1$-actions. Then there exists a constant $C > 0$ such that there are no nontrivial holomorphic $\mathbb{Z}_p$-actions of prime order on $X$ provided that $p > C$. Moreover, the constant $C$ depends linearly on the Betti numbers of $X$ and the order of the torsion subgroup of $H_2(X)$, i.e., there exists a universal constant $c > 0$ such that*

$$C = c(1 + b_1 + b_2 + |Tor\ H_2|).$$

A natural question is that to what extent Theorem 4.1 holds also in the symplectic category. Let $(X, \omega)$ be a symplectic 4-manifold such that $[\omega] \in H^2(X;\mathbb{Q})$. Denote by $N_\omega$ the smallest positive integer such that $[N_\omega \omega]$ is an integral class. We set

$$C_\omega \equiv N_\omega c_1(K) \cdot [\omega],$$

where $K$ is the canonical bundle of $(X, \omega)$.

**Theorem 4.2.** *(Chen [17]) Let $(X, \omega)$ be a symplectic 4-manifold with $b_2^+ > 1$ and $[\omega] \in H^2(X;\mathbb{Q})$, which does not satisfy the following condition: $X$ is minimal with vanishing Euler characteristic and signature. Then there exists a constant $C > 0$:*

$$C = c(1 + b_1^2 + b_2^2)C_\omega^2$$

*where $c > 0$ is a universal constant, such that there are no nontrivial symplectic $\mathbb{Z}_p$-actions of prime order on $X$ provided that $p > C$.*

A new phenomenon arising from Theorem 4.2 was that the constant $C$ may also depend on the smooth structure of the 4-manifold, as $c_1(K)$ is a Seiberg-Witten basic class and $C_\omega = N_\omega c_1(K) \cdot [\omega]$. Even though the following example does not say anything about the situation in Theorem 4.2, it does however suggest that the constant $C$ in Question 4.1 should be more than topological in nature.

**Example 4.1.** (Chen [17])

Let $X_0$ be the smooth rational elliptic surface given by the Weierstrass equation

$$y^2 z = x^3 + v^5 z^3.$$

For any prime number $p \geq 5$, one can define an order-$p$ automorphism $g$ of $X_0$ as follows (cf. [69]):

$$g : (x, y, z; v) \mapsto (\mu_p^{-5} x, y, \mu_p^{-15} z; \mu_p^6 v), \quad \mu_p \equiv \exp(2\pi i/p).$$

Note that $g$ preserves the elliptic fibration and leaves exactly the two singular fibers (at $v = 0$ and $v = \infty$) invariant.

Let $X_p$ be the symplectic 4-manifold obtained from $X_0$ by performing repeated knot surgery (cf. [30]) using the trefoil knot on $p$ copies of regular fibers of the elliptic fibration which are invariant under the order-$p$ automorphism $g$. Then it is clear that $X_p$ inherits a periodic symplectomorphism of order $p$. On the other hand, $X_p$ is homeomorphic to $X_0$. (The key point is that $X_p$ continues to be simply connected, as repeated knot surgery on parallel copies of a regular fiber is equivalent to a single knot surgery using the connected sum of the knots, see [17] for more details.)

To see that $X_p$ admits no smooth $\mathbb{S}^1$-actions, notice that the canonical class of $X_p$ is given by the formula

$$c_1(K_{X_p}) = (2p - 1) \cdot [T],$$

where $[T]$ is the fiber class of the elliptic fibration which pairs positively with the symplectic form $\omega$ on $X_p$. Since $c_1(K_{X_p}) \cdot [\omega] > 0$ and $c_1(K_{X_p})^2 = 0$, $X_p$ is neither rational nor ruled, cf. [45]. By Baldridge's theorem [4] or the strengthened version of the theorem of Fintushel in [28], $X_p$ does not admit any smooth $\mathbb{S}^1$-actions.

We thus obtained, for any prime number $p \geq 5$, a symplectic 4-manifold $X_p$ homeomorphic to the rational elliptic surface, which admits no smooth $\mathbb{S}^1$-actions but has a periodic symplectomorphism of order $p$. Observe that if $[\omega]$ is integral (this can be arranged), the order $p$ satisfies

$$p \leq \frac{1}{2}(c_1(K_{X_p}) \cdot [\omega] + 1).$$

Note that in Example 4.1, the manifold $X_p$ has $b_2^+ = 1$. It is not clear, however, that with the condition $b_2^+ > 1$, whether the dependence of the constant $C$ on $\omega$ (as well as the assumption $[\omega] \in H^2(X; \mathbb{Q})$) in Theorem 4.2 can be removed or not. Note that because of Theorem 4.1, the construction in Example 4.1 does not extend to the $b_2^+ > 1$ case by simply replacing the rational elliptic surface with some other elliptic surfaces with $b_2^+ > 1$.

The following is a symplectic analog of Xiao's theorem in [67].

**Question 4.2.** (Chen [17]) Let $(X, \omega)$ be a minimal symplectic 4-manifold of Kodaira dimension 2 (i.e., $c_1(K) \cdot [\omega] > 0$ and $c_1(K)^2 > 0$, cf. [45]). Does there exists a universal constant $c > 0$, such that

$$|G| \leq c \cdot c_1(K)^2$$

for any finite group $G$ of symplectomorphisms of $(X, \omega)$?

Note that the manifolds $X_p$ in Example 4.1 are minimal (cf. Gompf [34] and Usher [62]) with $c_1(K_{X_p})^2 = 0$, so they are not of Kodaira dimension 2 and do not give counterexamples.

**Question 4.3.** (Chen [17]) For the general case of smooth $\mathbb{Z}_p$-actions, is the constant $C$ in Question 4.1 "describable" in terms of invariants of the manifold? In particular, what invariants of the smooth structure (e.g. SW basic classes, SW invariants, etc.) will enter the constant and how do they enter the constant?

The existence of smooth $\mathbb{S}^1$-actions depends on the smooth structure in general (as shown by Example 4.1). However, by a theorem of Atiyah and Hirzebruch [3], a spin 4-manifold with non-zero signature does not admit smooth $\mathbb{S}^1$-actions for any given smooth structure. Example 4.1 naturally suggests the following question.

**Question 4.4.** (Chen [17]) Let $X$ be a simply connected smoothable 4-manifold with even intersection form and non-zero signature. Does there exist a constant $C > 0$ depending on the topological type of $X$ only, such that for any prime number $p > C$, there are no $\mathbb{Z}_p$-actions on $X$ which are smooth with respect to some smooth structure on $X$?

We end this survey with two results concerning homological rigidity of symplectic $\mathbb{Z}_p$-actions from [19]. The following theorem extends the homological rigidity of holomorphic actions on a $K3$ surface to the symplectic category.

**Theorem 4.3.** *(Chen-Kwasik* [19]*) There are no homologically trivial, symplectic actions of a finite group on the standard* $K3$ *surface (with respect to any symplectic structure).*

The following theorem partially extends Peters' results in [56] to the symplectic category.

**Theorem 4.4.** *(Chen-Kwasik* [19]*) Let* $X$ *be a minimal symplectic* 4*-manifold with* $c_1^2 = 0$ *and* $b_2^+ > 1$*, which admits a homologically trival (over* $\mathbb{Q}$*), pseudofree with* $X^G \neq \emptyset$*, symplectic* $\mathbb{Z}_p$*-action for a prime* $p > 1$*. Then the following conclusions hold.*

- (a) *The action is trivial if* $p \neq 1 \bmod 4$*,* $p \neq 1 \bmod 6$*. In particular, for infinitely many primes* $p$ *the manifold* $X$ *does not admit any such nontrivial* $\mathbb{Z}_p$*-actions.*
- (b) *The action is trivial as long as there is a fixed point* $x$ *of local representation* $\rho_x$ *contained in* $SL(2, \mathbb{C})$*.*

We remark that while the homological rigidity question aims at giving some sort of understanding toward Question 1.2 regarding complexity of the set of associated data $\Gamma = (\rho, X^G, \{\rho_x\})$ of all $\mathbb{Z}_p$-actions on a given 4-manifold $X$, proofs of Theorems 4.3 and 4.4 relied heavily on the knowledge of $(X^G, \{\rho_x\})$ which was obtained through the pseudoholomorphic curve techniques developed in [19].

Regarding homological rigidity of smooth actions, the following question remains open, and is perhaps the most interesting one.

**Question 4.5.** For any prime number $p > 2$, are there any nontrivial but homologically trivial, smooth $\mathbb{Z}_p$-actions on a homotopy $K3$ surface?

For the case of an involution, it is known that there are no locally linear, homologically trivial topological actions on a homotopy $K3$ surface (see [49, 57]). On the other hand,

Question 4.5 should be compared with Question 4.4, as for any $p > 23$, a locally linear $\mathbb{Z}_p$-action on a homotopy $K3$ surface is automatically homologically trivial by Prop 1.1(1).

**Acknowledgements**: Many of the results surveyed in this article were joint work with Sławomir Kwasik; I wish to thank him for the collaborations on this subject. Some of the ideas or open questions discussed here might occur or were formed during or after a conversation with him. The paper is an expanded version of the author's talk at the 16th Gökova Geometry/Topology Conference, Turkey, May 25-30, 2009. I wish to thank Selman Akbulut for encouraging me to write this survey.

## References

[1] S. Akbulut, *Cappell-Shaneson's 4-dimensional s-cobordism*, Geometry and Topology **6** (2002), 425–494.

[2] ——————, *Cappell-Shaneson homotopy spheres are standard*, Ann. of Math., to appear.

[3] M.F. Atiyah and F. Hirzebruch, *Spin-manifolds and group actions*, Essays on Topology and Related Topics (Mémoires dédiés à Georges de Rham), pp. 18–28, Springer, New York, 1970.

[4] S. Baldridge, *Seiberg-Witten vanishing theorem for $\mathbb{S}^1$-manifolds with fixed points*, Pacific J. of Math. **217** (2004), no.1, 1–10.

[5] ——————, *Seiberg-Witten invariants, orbifolds, and circle actions*, Tran. Amer. Math. Soc. **355** (2003), no.4, 1669–1697.

[6] ——————, *Seiberg-Witten invariants of 4-manifolds with free circle actions*, Commun. Contemp. Math. **3** (2001), 341–353.

[7] W. Barth, K. Hulek, C. Peters and A. Van de Ven, *Compact Complex Surfaces*, 2nd enlarged edition, Ergebnisse der Math. Vol. 4, Springer-Verlag, 2004.

[8] S. Bauer and D. Wilczynski, *On the topological classification of pseudofree group actions on 4-manifolds, II*, K-theory **10** (1996), no.5, 491–516.

[9] G. Bredon, *Introduction to Compact Transformation Groups*, Academic Press, New York, 1972.

[10] S.E. Cappell and J.L. Shaneson, *Some new four manifolds*, Ann. of Math. (2) **104** (1976), no.1, 61–72.

[11] ——————, *On 4-dimensional s-cobordisms*, J. Diff. Geom. **22** (1985), 97–115.

[12] ——————, *Smooth nontrivial 4-dimensional s-cobordisms*, Bull. Amer. Math. Soc. **17** (1987), no. 1, 141–143, *Corrigendum to: smooth nontrivial 4-dimensional s-cobordisms*, Bull. Amer. Math. Soc. **17** (1987), no. 2, 401.

[13] S.E. Cappell, J.L. Shaneson, M. Steinberger and J.E. West, *Non-linear similarity begins in dimension six*, Amer. J. Math. **111** (1989), 717–752.

[14] W. Chen, *Pseudoholomorphic curves in four-orbifolds and some applications*, in Geometry and Topology of Manifolds, Boden, H.U. et al ed., Fields Institute Communications **47**, pp. 11–37. Amer. Math. Soc., Providence, RI, 2005.

[15] ——————, *Orbifold adjunction formula and symplectic cobordisms between lens spaces*, Geometry and Topology **8** (2004), 701–734.

[16] ——————, *Smooth s-cobordisms of elliptic 3-manifolds*, Journal of Differential Geometry **73** no.3 (2006), 413–490.

[17] ——————, *On the orders of periodic diffeomorphisms of 4-manifolds*, preprint, 2010.

[18] ——————, *Some rigidity results of symplectic group actions*, unpublished, 2004.

[19] W. Chen and S. Kwasik, *Symplectic symmetries of 4-manifolds*, Topology **46** no.2 (2007), 103–128.

[20] ——————, *Symmetries and exotic smooth structures on a K3 surface*, Journal of Topology **1** (2008), 923–962.

[21] ——————, *Symmetric homotopy K3 surfaces*, arXiv:0709.2448 [math.GT].

[22] W. Chen, S. Kwasik, and D. Wilczynski, *Involutions of K3 surface with 8 fixed points*, in progress.

[23] A. Edmonds, *Construction of group actions on four-manifolds*, Trans. Amer. Math. Soc. **299** (1987), 155–170.
[24] ————, *Aspects of group actions on four-manifolds*, Topology and its Applications **31** (1989), 109–124.
[25] ————, *A survey of group actions on 4-manifolds*, arXiv:0907.0454 [math.GT].
[26] A. Edmonds and J. Ewing, *Locally linear group actions on the complex projective plane*, Topology **28** no.2 (1989), 211–223.
[27] ————, *Realizing forms and fixed point data in dimension four*, Amer. J. Math. **114** (1992), no.5, 1103–1126.
[28] R. Fintushel, *Classification of circle actions on 4-manifolds*, Tran. Amer. Math. Soc. **242** (1978), 377–390.
[29] R. Fintushel and R. Stern, *An exotic free involution on* $\mathbb{S}^4$, Ann. of Math. (2) **113** (1981), no.2, 357–365.
[30] ————, *Knots, links, and 4-manifolds*, Invent. Math.**134** (1998), 363–400.
[31] A. Garbagnati and A. Sarti, *Symplectic automorphisms of prime order on* $K3$ *surfaces*, J Algebra **318** (2007), no.1, 323–350.
[32] C.H. Giffen, *The generalized Smith conjecture*, Amer. J. Math. **88** (1966), 187–198.
[33] R.E. Gompf, *Killing the Akbulut-Kirby 4-sphere, with relevance to the Andrews-Curtis and Schoenflies problems*, Topology **30** (1991), no.1, 97–115.
[34] ————, *A new construction of symplectic manifolds*, Ann. of Math. (2) **142** (1995), no. 3, 527–595.
[35] C. McA. Gordon, *On the G-signature theorem in dimension four*, in A la Recherche de la Topologie Perdue, A. Marin and L. Guillou ed. Progr. Math. **62**, pp. 159–180, Birkhäuser, Boston, 1986.
[36] M. Gromov, *Pseudoholomorphic curves in symplectic manifolds*, Invent. Math. **82** (1985), 307–347.
[37] I. Hambleton and M. Kreck, *On the classification of topological 4-manifolds with finite fundamental group*, Math. Ann. **280** (1988), no.1, 85–104.
[38] I. Hambleton and R. Lee, *Smooth group actions on definite 4-manifolds and moduli spaces*, Duke Math. J. **78** (1995), no. 3, 715–732.
[39] I. Hambleton, R. Lee and I. Madsen, *Rigidity of certain finite group actions on the complex projective plane*, Comment. Math. Helvetici **64**(1989), 618–638.
[40] F. Hirzebruch and D. Zagier, *The Atiyah-Singer Theorem and Elementary Number Theory*, Math. Lecture Series **3**, Publish or Perish, Inc., 1974
[41] S. Kwasik and R. Schultz, *Homological properties of periodic homeomorphisms of 4-manifolds*, Duke Math. J. **58** (1989), 241–250.
[42] ————, *On s-cobordisms of metacyclic prism manifolds*, Invent. Math. **97** (1989), 523–552.
[43] ————, *Vanishing of Whitehead torsion in dimension four*, Topology **31** (1992), no. 4, 735–756.
[44] H.B. Lawson and S.T. Yau, *Scalar curvature, nonabelian group actions, and the degree of symmetry of exotic spheres*, Comment. Math. Helv. **49** (1974), 232–244.
[45] T.-J. Li, *The Kodaira dimension of symplectic 4-manifolds*, Clay Mathematics Proceedings, Volume 5, 2006, 249–261.
[46] X. Liu and N. Nakamura, *Pseudofree* $\mathbb{Z}/3$*-actions on* $K3$ *surfaces*, Proc. Amer. Math. Soc. **135** (2007), no. 3, 903–910.
[47] ————, *Nonsmoothable group actions on elliptic surfaces*, Topology Appl. **155** (2008), no. 9, 946–964.
[48] N. Machida and K. Oguiso, *On* $K3$ *surfaces admitting finite non-symplectic group actions*, J. Math. Sci. Univ. Tokyo **5**(1998), 273–297.
[49] T. Matumoto, *Homologically trivial smooth involutions on* $K3$ *surfaces*, in Aspects of low-dimensional manifolds, Matsumoto, Y. and Morita, S. ed., Advanced Studies in Pure Mathematics **20**, pp. 365–376, Kinokuniya, Tokyo, 1992.

[50] M. McCooey, *Symmetry groups of four-manifolds*, Topology **41** (2002), no.4, 835–851.
[51] D. McDuff and D. Salamon, *J-holomorphic Curves and Symplectic Topology*, Colloquium Publications **52**, AMS, 2004
[52] D.R. Morrison, *On K3 surfaces with large Picard number*, Invent. Math. **75** (1986), 105–121.
[53] S. Mukai *Finite groups of automorphisms of K3 surfaces and the Mathieu group*, Invent. Math. **94** (1988), 183–221.
[54] V.V. Nikulin, *Finite groups of automorphisms of Kählerian surfaces of type K3*, Trudy Mosk. Math. **38** (1979), 75–137, Trans. Moscow Math. Soc. **38** (1980), 71–135.
[55] ———, *Integral symmetric bilinear forms and some applications*, Izv. Math. Nauk SSSR **43** (1979), 111–177, Math. USSR Izvestija **14** (1980), 103–167.
[56] C.A.M. Peters, *On automorphisms of compact Kähler surfaces*, Algebraic Geometry Angers 1979, Alphen aan de Rijn: Sijthoff and Noordhoff 1980.
[57] D. Ruberman, *Involutions on spin 4-manifolds*, Proc. Amer. Math. Soc. **123**(1995), no.2, 593–596
[58] R. Schultz, *Transformation groups and exotic spheres*, in Group Actions on Manifolds, Schultz, R. ed., Contemporary Mathematics **36**, pp.243–267, Amer. Math. Soc., Providence, RI, 1985
[59] C.H. Taubes, *SW $\Rightarrow$ Gr: from the Seiberg-Witten equations to pseudoholomorphic curves*, J. Amer. Math. Soc. **9**, 845–918 (1996), and reprinted with errata in Proceedings of the First IP Lectures Series, Volume II, R. Wentworth ed., International Press, Somerville, MA, 2000
[60] ———, *The geometry of the Seiberg-Witten invariants*, Proceedings of ICM (Berlin, 1998), Volume II, 493–504
[61] K. Ueno, *A remark on automorphisms of Enriques surfaces*, J. Fac. Sci. Univ. Tokyo Sect. I A Math. **23** (1976), no. 1, 149–165.
[62] M. Usher, *Minimality and symplectic sums*, Int. Math. Res. Not. 2006, Art. ID 49857, 17 pp.
[63] D. Wilczynski, *Periodic maps on simply connected four-manifolds*, Topology **30** (1991), no.1, 55–65.
[64] ———, *On the topological classification of pseudofree group actions on 4-manifolds, I*, Math. Z. **217** (1994), no.3, 335–366.
[65] J.A. Wolf, *Spaces of Constant Curvature*, McGraw-Hill, New York, 1967.
[66] G. Xiao, *Galois covers between K3 surfaces*, Ann. Inst. Fourier (Grenoble) **46** (1996), 73–88.
[67] ———, *Bound of automorphisms of surfaces of general type, I*, Ann. of Math. (2) **139** (1994), no. 1, 51–77.
[68] ———, *Bound of automorphisms of surfaces of general type, II*, J. Algebraic Geom. **4** (1995), no. 4, 701–793.
[69] D.-Q. Zhang, *Automorphisms of finite order on Gorenstein Del Pezzo surfaces*, Trans. Amer. Math. Soc. **354** (2002), 4831–4845.

UNIVERSITY OF MASSACHUSETTS, AMHERST
*E-mail address*: `wchen@math.umass.edu`

Proceedings of $16^{th}$ Gökova
Geometry-Topology Conference
pp. 22 – 33

# Introduction to integrable systems: open Toda lattice, KP-, and KdV-hierarchies

*Michael Shapiro*

Abstract. The goal of this crash course is to make a brief introduction into the beautiful world of integrable hierarchies. We do not intend to give a general survey of integrable system theory but rather want to use few known examples to introduce notions and initial circle of ideas specific for integrable equations. We used [A] as a reference for the first section, [P] for the second section, and [MJD] and [K] for the third section.

## 1. Classical mechanics, Poisson structures, Hamiltonian formalism

The classical Newton's equations for a particle with coordinates $q(t) = (q_1(t), \dots, q_n(t))$ in a potential force field $\mathbf{F}$ have a form

$$\ddot{q} = \mathbf{F}, \ \mathbf{F} = -\operatorname{grad} P,$$

where $P = P(q)$ is a potential energy.

We rewrite these equations as

$$\begin{cases} \dot{q} & = p \\ \dot{p} & = -\operatorname{grad} P, \end{cases}$$

where $p = (p_1, \dots, p_n)$ is *the momentum*.

The space $\mathbb{R}^{2n} = \{(p, q)\}$ is *the phase space* of the system.

**Definition 1.1.** *A Poisson structure* on a manifold $M$ is a bilinear bracket $\{\cdot, \cdot\}$ on the space of functions $C^\infty(M) \times C^\infty(M) \to C^\infty(M)$ satisfying

(1) skew-symmetry $\{F, G\} = -\{G, F\}$
(2) Leibnitz rule $\{F, G \cdot H\} = \{F, G\} + \{F, H\}$
(3) Jacobi identity $\{\{F, G\}H\} + \{\{G, H\}F\} + \{\{H, F\}G\} = 0$

*Remark* 1.2. Leibnitz rule is equivalent to the following formula: in any local coordinate system $\{x_i\}$ on $M$ one can express the Poisson bracket as $\{F, G\} = \sum_i \dfrac{\partial F}{\partial x_i}\{x_i, G\}$.

---

*Key words and phrases.* integrable systems, Toda lattice, KP, KdV.
Supported by TÜBİTAK .

**Example 1.3.** The standard Poisson structure on $\mathbb{R}^{2n} = \{(p, q)\}$ is given by

$$\{F, G\} = \sum_i \frac{\partial F}{\partial q_i}\frac{\partial G}{\partial p_i} - \frac{\partial F}{\partial p_i}\frac{\partial G}{\partial q_i}$$

This Poisson structure corresponds to a bi-vector field

$$\nu = \sum_i \frac{\partial}{\partial q_i} \wedge \frac{\partial}{\partial p_i},$$

such that $\{F, G\}(x) = \langle dF(x) \wedge dG(x), \nu(x)\rangle$, where $\langle *, *\rangle$ is the natural pairing between $\Lambda^2 T_x^*(X)$ and $\Lambda^2 T_x(X)$.

A Poisson bracket transforms a covector field into a vector field, more exactly it induces a linear map $\Psi_x : T_x^* M \to T_x M$. For any two covectors $\xi, \eta \in T_x^* M$ let $\xi = dF(x)$, $\eta = dG(x)$. Then $\Psi_x$ is determined uniquely from the equality $\eta(\Psi_x \xi) = \{F, G\}(x)$. Hence, a Poisson bracket transforms any function $H$ into a vector field $\operatorname{sgrad} H = \Psi_x(dH(x))$.

## 1.1. Hamiltonian equation

Given *a hamiltonian function (or hamiltonian)* $H(x)$ we define the Hamiltonian flow (or Hamiltonian equation) as

$$\dot{x} = \{x, H\} \tag{1.1}$$

For $x = (p, q)$ we have $H = H(p, q), \dot{p} = \{p, H\}, \dot{q} = \{q, H\}$.

*Remark* 1.4. For any function $f(x)$ we have $\dot{f} = \sum_i \frac{\partial f}{\partial x_i}\dot{x}_i = \sum_i \frac{\partial f}{\partial x_i}\{x_i, H\} = \{f, H\}$ by the Leibnitz rule.

**Example 1.5.** Consider $x = (p, q) \in \mathbb{R}^{2n}$, equipped with the standard Poisson structure $\{F, G\} = \sum_i \frac{\partial F}{\partial q_i}\frac{\partial G}{\partial p_i} - \frac{\partial F}{\partial p_i}\frac{\partial G}{\partial q_i}$, and the hamiltonian $H(p, q) = \frac{p^2}{2} + P(q)$ (the total energy). Then the flow equations take the form

$$\begin{cases} \dot{q} = p \\ \dot{p} = -\operatorname{grad} P \end{cases}$$

We recognize Newton's equations of motion.

## 1.2. Integrals of motion

Let $I(x)$ be a function such that $\{I, H\} = 0$. Then $I$ is preserved under the Hamiltonian flow with hamiltonian $H$. Indeed, $\dot{I} = \{I, H\} = 0$ and $I(t) = I(p(t), q(t)) = Constant$.

**Definition 1.6.** Such $I$ is called *a first integral of motion.* A collection $I_1, \ldots, I_k$ of the first integrals such that $\{I_j, I_l\} = 0$ for all $j$ and $l$ form *integrals in involution.* Integrals are called *independent* if their gradients are linearly independent at a generic point of $M$.

## 1.3. Liouville's theorem

A $2n$ dimensional Poisson manifold $M^{2n}$ is *symplectic* if the bivector field $\nu$ giving the Poisson structure is nondegenerate at every point of $M$.

*Theorem* 1.7. Let a symplectic $M^{2n}$ have $n$ functions $F_1, \ldots, F_n$ in involution, i.e., $\{F_i, F_j\} \equiv 0$, for all $i, j$. Consider $M_f := \{x | F_i(x) = f_i, i = 1 \ldots n\}$. Assume that all $F_i$ are independent on $M_f$ (i.e. $dF_i(x)$ are linearly independent for all $x \in M_f$). Then,

(1) $M_f$ is a smooth manifold invariant with respect to hamiltonian flow $\operatorname{sgrad} F_1$.
(2) If $M_f$ is compact and connected then

$$M_f \approx T^n = \{(\varphi_1, \ldots, \varphi_n) mod 2\pi\}.$$

(3) The flow in the phase space with hamiltonian $H = F_1$ determines on $M_f$ a quasi-periodic motion in angle coordinates $\varphi_1, \ldots, \varphi_n$. Namely, $\dot{\varphi}_i = \omega_i$, where $\omega = (\omega_1, \ldots, \omega_n)$ depends on $f$.
(4) Hamiltonian equations with hamiltonian $H = F_1$ are solved in quadratures (i.e. there are analytic formulas for the motion).

For the system with $n$ degrees of freedom (dimension of the phase space is $2n$) there are at most $n$ independent integrals in involution $H_1, \ldots, H_n$. They are the *action variables*. If $H = H_1$ is a hamiltonian then there are $n$ additional coordinates (called *angle variables*) $\varphi_1, \ldots, \varphi_n$ such that $0 = \{\varphi_i, \varphi_j\} = \{H_i, H_j\}$, $\{\varphi_i, H_j\} = 0$ for $i \neq j$, $\{\varphi_i, H_i\} = \delta_{ij}$.
The solution of the Cauchy problem with initial condition $(H_j(0), \varphi_j(0))$ is

$$\begin{aligned} H_j(t) &= H_j(0) \\ \varphi_j(t) &= \varphi_j(0) + H_j(0)t. \end{aligned}$$

# 2. Open Toda lattice

We consider the evolution of $n$ labeled interacting particles on a line where only $i$th and $i+1$st particles interact. The corresponding hamiltonian is given by

$$H = \frac{1}{2} \sum_{j=1}^{n} p_j^2 + g^2 \sum_{j=1}^{n-1} exp(2(q_j - q_{j+1})),$$

where $q_j$ is the coordinate of $j$th particle, $p_j = \dot{q}_j$ is its momentum. The Poisson structure in coordinates $(p_i, q_i)$ is standard.

By the change of variables $q_j \to q_j + a_j$ we can always make $g = 1$. The corresponding differential equations take the form

$$\begin{cases} \dot{q_1} &= p_1 \\ \dots & \\ \dot{q_n} &= p_n \\ \dot{p_1} &= -2e^{2(q_1-q_2)} \\ \dot{p_2} &= -2e^{2(q_2-q_3)} + 2e^{2(q_1-q_2)} \\ \dots & \\ \dot{p_n} &= 2e^{2(q_{n-1}-q_n)} \end{cases} \tag{2.1}$$

Note that the system is invariant under the shift $q_i \to q_i + a$, and therefore we can reduce the dimension of the system by 1. Moreover, recall that the total momentum of the closed system is preserved. Without loss of generality we will assume that $\sum_i p_i = 0$ so that the dimension of the system becomes $2n - 2$.

To solve the system we want to find integrals in involution. To find the integrals we look for the Lax form of 2.1. Namely, we are looking for a way to construct the matrix $L$ such that matrix Lax equation

$$\dot{L} = [L, L_+] \tag{2.2}$$

is equivalent to 2.1. Then the coefficients of the characteristic polynomial of $L$ give integrals of motion. Such a matrix $L = L(p_i, q_i)$ can be found in a tridiagonal form,

$$L = \begin{pmatrix} p_1 & e^{q_1-q_2} & 0 & \dots & \\ e^{q_1-q_2} & p_2 & e^{q_2-q_3} & \dots & \\ & & \dots & & \\ & & & \dots & e^{q_{n-1}-q_n} \\ & & \dots & e^{q_{n-1}-q_n} & p_n \end{pmatrix} \tag{2.3}$$

and $L_+$ is a skew-symmetrization of $L$

$$L_+ = \begin{pmatrix} 0 & e^{q_1-q_2} & 0 & \dots & \\ -e^{q_1-q_2} & 0 & e^{q_2-q_3} & \dots & \\ & & \dots & & \\ & & & \dots & e^{q_{n-1}-q_n} \\ & & \dots & -e^{q_{n-1}-q_n} & 0 \end{pmatrix} \tag{2.4}$$

*Home Assignment: Check that Lax equation 2.2 is equivalent to 2.1.*

Denote $a_i = exp(q_i - q_{i+1})$ for $i = [1, n-1]$, $b_j = p_j$ for $j = [1, n]$. As above we have $\sum_i b_i = 0$.

The flow in coordinates of $a_i, b_i$ is

$$\begin{cases} \dot{a}_k &= a_k(b_k - b_{k+1}) \\ \dot{b}_k &= 2(-a_k^2 + a_{k-1}^2) \end{cases} \tag{2.5}$$

The Poisson structure in coordinates $a_i, b_i$ has the form $\{b_j, a_{j-1}\} = -a_{j-1}$, and $\{b_j, a_j\} = a_j$, while all other brackets are zeroes.

*Remark* 2.1. This Poisson bracket is a particular case of Kirillov-Konstant Poisson structure on an orbit of a coadjoint action by the Borel subgroup $B$ of $SL_n$ on the dual space $b^*$ of the Borel subalgebra $b \subset sl_n$.

Here,

- $B$ is the subgroup of the non-degenerate upper triangular matrices.
- $\mathfrak{b}$ is the subalgebra of the upper-triangular matrices.
- $\langle X, Y\rangle = Const \cdot Tr(XY)$ is the non-degenerate symmetric Killing form.
- We identify $\mathfrak{b}^*$ with the space $Sym_n$ of symmetric $n \times n$ matrices with zero trace, it is dual to $\mathfrak{b}$ with respect to the Killing form $\langle *, *\rangle$.

Co-adjoint action of $B$ on $\mathfrak{b}^*$ is given by the following formula. For $X \in B$ and $y \in \mathfrak{b}^*$, we have $ad^*_X(y) = \mathrm{sym}\,(XyX^{-1})$, where sym $: sl_n \to Sym_n$ is the projection onto $Sym_n$ along $\mathfrak{b}$.

The Kirillov-Konstant Poisson bracket is defined as follows. For any $f, g \in C^\infty(b^*; \mathbb{R})$ we define $\{f, g\}_{KK}(x) = \langle x, [df(x), dg(x)]\rangle$, where $[df(x), dg(x)]$ is the Lie bracket of $df(x)$ and $dg(x)$ in the Lie algebra $\mathfrak{b}$ (note that both $df(x), dg(x) \in \mathfrak{b}$). Orbits of $ad^*_B$ are symplectic leaves of $\{,\}_{KK}$ in $\mathfrak{b}^*$.

*Symplectic leaf* of a Poisson manifold $M$ is an equivalence class of points in $M$ such that one can reach from one point to another in the same class along piecewise trajectories of hamiltonian flows. Tridiagonal matrices form an orbit, i.e. a symplectic leaf in $\mathfrak{b}^* \simeq Sym_n$. The Kirillov-Kostant Poisson bracket on the space of tridiagonal matrices coincides with the Toda lattice Poisson bracket written in coordinates $a_i, b_i$. Hamiltonian $H(x) = \frac{1}{2}Tr(x^2)$ induces the hamiltonian flow which coincides with the open Toda lattice flow.

*Home Assignment: check that hamiltonian flow on tridiagonal matrices equipped with Kirillliov-Konstant Poisson bracket with hamiltonian $H(x) = \frac{1}{2}Tr(x^2)$ is given by 2.5.*

*Remark* 2.2. The Cauchy problem $\dot{L} = [L, L_+]$, $L(0) = L_0$ has a solution of the form $L(t) = u(t)L_0u(t)^{-1}$, where $u(t)$ is an orthogonal matrix.

*Remark* 2.3. The characteristic polynomial is preserved under Toda flow.

*Remark* 2.4. Functions $H_k(L) = \frac{1}{k}Tr(L^k)$, $k = [1, n]$ are integrals of motion.

Let $det(\lambda I - L) = \prod_{k=1}^n(\lambda - \lambda_k) = \sum_{k=1}^n(-1)^k J_k(L)\lambda^{n-k}$.

*Remark* 2.5. Functions $J_k(L), k = [1, n]$ are integrals of motion.

*Remark* 2.6. Functions $\lambda_k(L), k = [1, n]$ are integrals of motion.

Now we show that restrictions of functions invariant under $sl_n$-conjugation Poisson commute with each other.

Let $f$ be a functions on $sl_n$, and let $\tilde{f}$ denote its restriction $f|_{\mathfrak{b}^*} = f|_{Sym_n}$. Recall that $sl_n = \mathfrak{b} \oplus \mathfrak{n}$, where $\mathfrak{n}$ denotes the Lie subalgebra of skew-symmetric matrices, $\mathfrak{n} = Sym_n^{\perp} = (\mathfrak{b}^*)^{\perp}$ with respect to the Killing form. Note that $d\tilde{f} = \pi_{\mathfrak{b}} f$, where $\pi_{\mathfrak{b}}$ is the orthogonal projection onto $\mathfrak{b}$. If $f$ and $g$ are invariant under conjugation then $\langle df(x), [x,u]\rangle = \langle dg(x), [x,u]\rangle = 0$ for any $x \in \mathfrak{b}, u \in sl_n$. Then $\{\tilde{f},\tilde{g}\}(x) = \langle x, [\pi_{\mathfrak{b}}(df(x)), \pi_{\mathfrak{b}}(dg(x))]\rangle$. Rewrite the latter using $\pi_{\mathfrak{b}}(df(x)) = df(x) - \pi_{\mathfrak{n}}(df(x))$, and similarly for $g$. More exactly,

$$\begin{aligned}\langle x, [\pi_{\mathfrak{b}}(df(x)), \pi_{\mathfrak{b}}(dg(x))]\rangle &= \langle x, [df(x) - \pi_{\mathfrak{n}}(df(x)), dg(x) - \pi_{\mathfrak{n}}(dg(x))]\rangle \\ &= \langle x, [\pi_{\mathfrak{n}}(df(x)), \pi_{\mathfrak{n}}(dg(x))]\rangle = 0\end{aligned}$$

since $[\pi_{\mathfrak{n}}(df(x)), \pi_{\mathfrak{n}}(dg(x))] \in \mathfrak{n} = (\mathfrak{b}^*)^{\perp}$.

*Remark* 2.7. Three collections $\{H_k\}$, $\{J_k\}$, $\{\lambda_k\}$ are collections of integrals in involution.

## 2.1. Solution of Toda lattice

We found above three different complete sets of integrals in involution. We can choose any set as a collection of action variables and find the corresponding angle variables $\varphi_k$. This will lead to a complete solution of the corresponding differential equations. However, there is a simpler method to solve Toda equations that we are going to discuss now.

### 2.1.1. Factorization method

As we saw above we can look for a solution in the form

$$L(t) = u(t)L_0u(t)^{-1}, \; u(0) = I. \tag{2.6}$$

In order to find $L(t)$ it is enough to compute $u(t)$.

Factorize $exp(L_0 \cdot t)$ into the product

$$exp(L_0t) = b(t) \cdot u(t), \tag{2.7}$$

where $u(t) \in SO(n)$ is orthogonal, $b(t) \in B$ is upper-triangular. Then substituting $u(t)$ into 2.6 gives the solution.

*Proof.* Let $L(t) = u(t)L_0u(t)^{-1}$. Its derivative is $\dot{L} = \dot{u}L_0u(t)^{-1} - u(t)L_0u(t)^{-1}\dot{u}(t)u(t)^{-1} = \dot{u}(t)u(t)^{-1} \cdot u(t)L_0u(t)^{-1} - u(t)L_0u(t)^{-1} \cdot \dot{u}(t)u(t)^{-1} = \left[\dot{u}(t)u(t)^{-1}, L(t)\right]$.

It remains to show that

$$\dot{u}(t)u(t)^{-1} = \pi_+(u(t)L_0u(t)^{-1}),$$

where $\pi_+$ is the projection along the upper triangular matrices onto the subspace of skew-symmetric matrices. Let $exp(L_0t) = b(t)u(t)$. Then $\dot{b}u + b\dot{u} = exp(L_0t)L_0 = b(t)u(t)L_0$. Hence, $b^{-1}\dot{b} + \dot{u}u^{-1} = uL_0u^{-1} = L(t)$. Note that, $\dot{u}u^{-1}$ is skew-symmetric, while $b^{-1}\dot{b}$ is upper-triangular. This observation accomplishes the proof. □

**Corollary 2.1.** $q_k(t) = q_k(0) + \frac{1}{2}\ln\frac{\Delta_{n+1-k}(t)}{\Delta_{n-k}(t)}$, *where* $\Delta_j$ *is the determinant of the lower right* $j \times j$ *submatrix of* $exp(2L_0t)$.

### 2.1.2. Moment method

Let $f(\lambda) = R_{nn}(\lambda) = \left((\lambda I - L)^{-1}\right)_{nn}$.

$$f(\lambda) = \sum_{i=1}^{n} \frac{r_k^2}{\lambda - \lambda_k},$$

where $r_k = u_n^{(k)}$, $u^{(k)} = (u_1^{(k)}, \dots, u_n^{(k)})$ is an eigenvector of $L$, $Lu^{(k)} = \lambda_k u^{(k)}$, and $(u^{(k)}, u^{(k)}) = 1$, $u_n^{(k)} > 0$.

*Remark* 2.8. $\sum_{k=1}^{n} r_k^2 = 1$

First we compute $r_k(t)$. Functions $r_k(t)$ satisfy differential equations.

Let $L(t) = u(t)\Lambda u(t)^{-1}$, where $\Lambda = diag(\lambda_1, \dots, \lambda_n)$. We show that $\dot{u^{(k)}} = -Mu^{(k)}$, where $M = L_+$. Then $\dot{L} = \dot{u}\Lambda u^{-1} - u\Lambda u^{-1} \cdot \dot{u}u^{-1} = [\dot{u}u^{-1}, L]$. Hence, $\dot{u}u^{-1} = M$, and $\dot{u} = Mu$.

Then, $\dot{r_k} = \dot{u}_n^{(k)} = -\left(Mu^{(k)}\right)_n = a_{n-1}u_{n-1}^{(k)}$.

On the other hand, $Lu^{(k)} = \lambda_k u^{(k)}$.

$a_{n-1}u_{n-1}^{(k)} = (\lambda_k - b_n) r_k$,

$b_n = L_{nn} = \sum_{k=1}^{n} \lambda_k r_k^2$.

Then, $\dot{r_k} = (\lambda_k - \sum_j \lambda_j r_j^2) r_k$, $\lambda_1 < \cdots < \lambda_n$.

Solution to this equation is $r_k^2(t) = \dfrac{r_k^2(0)e^{2\lambda_k t}}{\sum_{j=1}^{n} r_j^2(0)e^{2\lambda_j t}}$. To restore the solution of Toda lattice it remains to recover $a_j$ and $b_j$ in terms of $\lambda_j$, $r_k^2$. Here we follow Moser's procedure of continuous fractions. Define,

$$f(\lambda) = f_n(\lambda) = \cfrac{1}{\lambda - b_n - \cfrac{a_{n-1}^2}{\lambda - b_{n-1} - \cfrac{a_{n-2}^2}{\lambda - b_{n-2} - \cfrac{\cdots}{\cdots - \cfrac{a_1^2}{\lambda - b_1}}}}}.$$

Let $\Delta_k$ be the upper left $k \times k$ minor of $(\lambda I - L)$. Then, $\Delta_k = (\lambda - b_k)\Delta_{k-1} - a_{k-1}^2 \Delta_{k-2}$. Set $s_k = \frac{\Delta_k}{\Delta_{k-1}}$, $s_1 = \Delta_1$, $\Delta_0 = 1$. By induction, $s_k = (\lambda - b_k) - a_{k-1}^2 / s_{k-1}$. On the other hand, Moser showed that $s_k = f_k^{-1}$, where $f_k = f|_{k \times k \text{ upper left submatrix of } (\lambda I - L)}$.

Then, $f_k = \dfrac{1}{\lambda - b_k - a_{k-1}^2 f_{k-1}}$.

This approach allows one to obtain formulas

$$b_l(t) = \frac{d}{dt} \ln\left(\frac{Q_l(t)}{Q_{l-1}(t)}\right), \; a_l(t) = \frac{\sqrt{Q_{l-1}(t)Q_{l+1}(t)}}{Q_l(t)},$$

where $Q_l(t)$ is the principal $l \times l$ minor of the Hankel matrix $H(t) = (h_{j+k}(t))_{j,k=1}^n$ satisfying the linear system

$$\dot{H}(t) = \begin{pmatrix} 0 & 1 & & 0 \\ \vdots & & \ddots & \\ 0 & & & 1 \\ -p_0 & -p_1 & \dots & -p_{n-1} \end{pmatrix} \cdot H(t)$$

with initial data $h_i(0) = (L^i(0))_{00} (i = 1, \dots, 2n)$ and coefficients $p_0, \dots, p_{n-1}$ defined by $det(L - \lambda I) = \lambda^n + \sum_{j=0}^{n-1} p_j \lambda^j$.

In particular, for $n = 2$ we get $b_1 = \dfrac{\lambda_2 r_1^2 + \lambda_1 r_2^2}{r_1^2 + r_2^2}$, $b_2 = \dfrac{\lambda_1 r_1^2 + \lambda_2 r_2^2}{r_1^2 + r_2^2}$, $a_1 = \dfrac{(\lambda_2 - \lambda_1) r_1 r_2}{r_1^2 + r_2^2}$.

## 3. KP-, and KdV-hierarchies

**1.** KdV-equation $u_t = -3uu' - au'''$, where $u = u(t, x)$ is considered as a time evolution for function $u(x)$. We want to describe Hamiltonian formalism and Lax form for KdV.

We are looking for integrals in involutions. Integrals in involution induce commuting flows. Such system of commuting flows is called *an integrable hierarchy.*

**2.** KP-hierarchy (KdV is a specialization of KP-hierarchy)

Let $F(p_1, p_2, \dots)$ be a function of infinite number of variables $p_1, p_2, \dots$. KP-hierarchy is the collection of differential equations:

$$\begin{cases} \dfrac{3}{4}\dfrac{\partial^2 F}{\partial p_2^2} = \dfrac{\partial}{\partial p_1}\left(\dfrac{\partial F}{\partial p_3} - \dfrac{3}{2} F \dfrac{\partial F}{\partial p_1} - \dfrac{1}{4}\dfrac{\partial^3 F}{\partial p_1^3}\right) \\ \qquad \dots \end{cases} \tag{3.1}$$

Below we discuss how to obtain these partial differential equations in terms of hamiltonian flows on the space of pseudo-differential symbols.

**Definition 3.1.** We define *a pseudo-differential symbol* $A := \sum_{k=-N}^{\infty} a_k \partial^{-k}$, $(\partial = \frac{d}{dx}, a_i(x) \in C^\infty(S^1))$, $deg(A) = N$ if $a_N(x) \not\equiv 0$, and $a_n(x) \equiv 0 \ \forall n > N$. Its *differential part* is $A_+ = \sum_{k=-N}^{0} a_k \partial^{-k}$.

Pseudo-differential symbols form a noncommutative associative algebra with multiplication defined as composition of operators using the following commutation relation $\partial^n \circ f = \sum_{j \geq 0} \binom{n}{j} (\partial^j f) \circ \partial^{n-j}$, where $\binom{n}{j} = \frac{n(n-1)\dots(n-j+1)}{j(j-1)\dots 1}$ defined for all $n \in \mathbb{Z}$ and $j \in \mathbb{Z}_{>0}$.

**Example 3.2.** $S = \partial^2 + u$ is a Schrödinger operator. Then $L = S^{\frac{1}{2}} = \partial + \sum_{n=1}^{\infty} f_n \partial^{-n}$. The Coefficients $f_n$ can be determined recurrently to give

$$L = \partial + \frac{1}{2} u \partial^{-1} - \frac{1}{4} u_x \partial^{-2} + \left(\frac{u_{xx}}{8} - \frac{u^2}{8}\right) \partial^{-3} + \dots$$

**Definition 3.3.** $\Psi$ is the Lie algebra of pseudo-differential operators. The commutator is defined by $[A, B] = A \circ B - B \circ A$.

**Definition 3.4.** Residue $res : \Psi \to C^\infty(S^1)$, is given by$(\sum a_i\partial^{-i}) \mapsto a_{-1}(x)$.

**Proposition 3.1.** $res([A,B]) = 0,\ \forall A, B \in \Psi$

**Definition 3.5.** Let $Tr(A) = \int_{S^1} res(A)$, for $A \in \Psi$. This gives rise to a symmetric bilinear form on $\Psi$: $(A,B) = Tr(AB) = \int_{S^1} res(A \circ B)$.

We introduce the special Lie subalgebra $\Psi_{int} = \{\partial + \sum_{k=1}^{\infty} a_k(x)\partial^{-k}\}$.

There is a natural *Gelfand-Dikii Poisson structure* on $\Psi_{int}$. Let $L \in \Psi_{int}$. Note that tangent vector to $\Psi_{int}$ has a form $\delta L = \sum_{k=-1}^{\infty} \delta a_k(x)\partial^{-k}$, $T_L\Psi_{int} = \{\sum_{k=-1}^{\infty} \delta a_k(x)\partial^{-k}\}$.

A cotangent vector is a differential operator $A = \sum_{m=1}^{N} a_m(x)\partial^m$. It determines the linear functional $F_A(\delta L) = \langle A, \delta L\rangle = Tr(\delta L \circ A)$.

It is enough to define the Poisson bracket on linear functionals as follows, then we can extend the Poisson bracket to other functions by the Leibniz rule.

$$\{F_A, F_B\}(L) = F_B\left((LA)_+ L - L(AL)_+\right).$$

Assume now that a pseudo-differential $L \in \Psi_{int}$ depends on $p_2, p_3, \ldots$ For a hamiltonian $H_k = \frac{1}{k} Tr(L^k)$, $k \geq 2$ the corresponding Hamiltonian equation has the form

$$\frac{\partial L}{\partial p_k} = [(L^k)_+, L], \tag{3.2}$$

where $X_+$ is pure differential part of $X$. These equations form KP-hierarchy for $L$. They are related by the so called $\tau$-function and it will be commented below.

### 3.1. Reduction of KP to KdV

Let $S = \partial^2 + u$ satisfy equations

$$\frac{\partial S}{\partial p_k} = [(S^{\frac{k}{2}})_+, S]$$

*Remark* 3.6. If $S$ is a differential operator (DO) then $S^{\frac{k}{2}}$ is also a DO for even $k$. Then $(S^{\frac{k}{2}})_+ = S^{\frac{k}{2}}$, and $[(S^{\frac{k}{2}})_+, S] = 0$ for even $k$. Hence, $S$ does not depend on $p_k$ for even $k$.

All coefficients of $L = S^{\frac{1}{2}}$ are determined just by one function $u$ (see Example 3.2).

KdV equation can be written in Lax form as

$$\dot{S} = [\left(S^{3/2}\right)_+, S] \tag{3.3}$$

Any solution of KP that does not depend on even $p_{2k}$ is a solution of KdV.

### 3.2. $\tau$-function

KP-hierarchy 3.2 describing commuting flows of a pseudo-differential operator $L$ is an infinite collection of equations written for an infinite number of unknown functions, namely the coefficients of the pseudo-differential operator $L$. However, one can rewrite all these equations in terms of a unique unknown function. This is called the *$\tau$-function.* Let $L = \partial + \sum_{j=1}^{\infty} u_i\partial^{-i} \in \Psi_{int}$ be a pseudo-differential operator of order 1.

Let $\mathbf{p}$ denote the infinite collection of variables $\mathbf{p} = (p_1, p_2, \dots)$ where $p_1 = x$. Through the dressing transformation $L = W\partial W^{-1}$ with $W = 1 + \sum_{i=1}^{\infty} w_i \partial^{-i}$ we obtain the equivalent Sato form of the hierarchy

$$\frac{\partial W}{\partial p_n} = (L^n)_+ W - W\partial^n. \tag{3.4}$$

The system 3.2 is also equivalent to the following eigenvalue problem

$$Lw = kw, \tag{3.5}$$

where $w = e^{\xi(\mathbf{p},k)}\left(1 + \frac{w_1}{k} + \frac{w_2}{k^2} + \dots\right)$, $\xi(\mathbf{p}, k) = \sum_{j=1}^{\infty} p_j k^j$. Note that $\frac{\partial}{\partial p_j} e^{\xi(\mathbf{p},k)} = k^j e^{\xi(\mathbf{p},k)}$. Consider the system of linear equations

$$\frac{\partial w}{\partial p_j} = (L^j)_+ w. \tag{3.6}$$

System 3.2 is the compatibility condition for 3.5 and 3.6.

The function $w$ can be expressed in terms of a function $\tau$ as

$$w = \frac{\tau\left(p_1 - \frac{1}{k}, p_2 - \frac{1}{2k^2}, \dots\right)}{\tau(p_1, p_2, \dots)} e^{\xi(\mathbf{p},k)} \tag{3.7}$$

The constants $w_1, w_2, \dots$ can be written in terms of $\tau$ as follows,

$$\begin{aligned} w_1 &= -\frac{\partial \tau}{\partial p_1} \cdot \tau^{-1} \\ w_2 &= \frac{1}{2}\left(\frac{\partial^2 \tau}{\partial p_1^2} - \frac{\partial^2 \tau}{\partial p_2^2}\right) \cdot \tau^{-1}. \end{aligned}$$

So, one can consider KP-hierarchy as an infinite system of equations for one function $\tau$.

*Remark* 3.7. The function $u$ in the KdV-equation is $u = 2\frac{\partial^2}{\partial p_1^2} \log \tau$.

*Remark* 3.8. If $F$ is a solution of KP system (3) then $e^F$ is a $\tau$-function.

*Remark* 3.9. KP-system 3.2 is equivalent to one Hirota bilinear difference equation on $\tau$-function

$$\begin{aligned} 0 = (\lambda_2 - \lambda_1)(\lambda_4 - \lambda_3)\tau(\mathbf{p} + [\lambda_1] + [\lambda_2])\tau(\mathbf{p} + [\lambda_3] + [\lambda_4]) - \\ (\lambda_4 - \lambda_2)(\lambda_3 - \lambda_1)\tau(\mathbf{p} + [\lambda_1] + [\lambda_3])\tau(\mathbf{p} + [\lambda_2] + [\lambda_4]) + \\ (\lambda_4 - \lambda_1)(\lambda_3 - \lambda_2)\tau(\mathbf{p} + [\lambda_1] + [\lambda_4])\tau(\mathbf{p} + [\lambda_2] + [\lambda_3]) \end{aligned}$$

where the *Miwa shift* is defined as

$$\mathbf{p} + [\lambda] = \left(p_1 + \lambda, p_2 + \frac{\lambda^2}{2}, p_3 + \frac{\lambda^3}{3}, \dots\right).$$

We want to describe all solutions of KP, (or, equivalently, all $\tau$-functions). Note that $\tau \equiv 1$ is a $\tau$-function. We obtain all other $\tau$-functions by a certain "Lie group" action. More exactly, the algebra $\widehat{gl}(\infty)$ (a certain extension of $gl(\infty)$ described below) acts on the space of solutions of KP, the "corresponding Lie group $\widehat{GL}(\infty)$" acts on $\tau$-functions and this action is transitive. In particular, any $\tau$-function lies in the orbit of $\tau \equiv 1$. To describe all $\tau$-functions we discuss below *boson-fermion correspondence* with applications to Sato Grassmannian and the KP-hierarchy.

**Schur polynomials.** We define elementary Schur polynomials $S_k(\mathbf{p})$ whose generating function is

$$\exp\left(\sum_{k=1}^{\infty} p_k z^k\right) = \sum_{k=0}^{\infty} S_k(\mathbf{p}) z^k$$

For $\lambda = (\lambda_1 > \lambda_2 > \cdots > \lambda_l)$, $\lambda_i \in \mathbb{Z}_{\geq 0}$ we define the Schur polynomial to be $S_\lambda(\mathbf{p})) = det\,(S_{\lambda_i - i + j}(\mathbf{p}))$.

Let $\bar{y} = y_1, \ldots, y_N$. The Schur function is defined by

$$s_\lambda(\bar{y}) = \frac{det\left(y_j^{\lambda_i + N - i}\right)_{i,j=1}^{N}}{det\left(y_j^{N-i}\right)_{i,j=1}^{N}} = det(h_{\lambda_i + j - i})_{i,j=1}^{N},$$

where $h_k = \sum_{1 \leq i_1 \leq i_2 \leq \cdots \leq i_k \leq N} y_{i_1} \cdots y_{i_k}$ is the complete homogeneous symmetric polynomial in $y_i$. Define $s_\lambda(y_1, y_2, \ldots) = \lim_{N \to \infty} s_\lambda(\bar{y})$.

Set $p_k = \frac{1}{k} \sum_{i=1}^{N} y_i^k$. Then $S_\lambda(\mathbf{p}) = s_\lambda(y_1, y_2, \ldots)$.

**Example 3.10.** $S_0 = 1$, $S_1 = p_1$, $S_2 = \frac{1}{2}p_1^2 + p_2$, $S_{1,1} = \frac{1}{2}p_1^2 - p_2$, $S_{2,1} = \frac{1}{3}p_1^3 - p_3, \ldots$

Note that Schur polynomials form a basis in $\mathbb{C}[[\mathbf{p}]]$, i.e. $\mathbb{C}[[\mathbf{p}]] \simeq \overline{\oplus_\lambda \mathbb{C} S_\lambda(\mathbf{p})}$.

Let $V = \mathbb{C}[z^{-1}][[z]]$ be the space of formal Laurent series.

Let $\Lambda^{\frac{\infty}{2}} V = span\{v_\lambda := z^{k_1} \wedge z^{k_2} \wedge \ldots\}$, where $k_i = \lambda_i - i$. Note that $k_i = -i$ for sufficiently large $i$.

*Remark* 3.11. $\mathbb{C}[[p_1, p_2, \ldots]]$ is the bosonic Fock space, $\Lambda^{\frac{\infty}{2}} V$ is the fermionic Fock space.

*Boson-fermion correspondence*: $\mathbb{C}[[p_1, p_2, \ldots]] \simeq \Lambda^{\frac{\infty}{2}} V$, $S_\lambda \leftrightarrow v_\lambda$.

*Remark* 3.12. Vacuum vector $v_\emptyset = z^{-1} \wedge z^{-2} \wedge z^{-3} \wedge \ldots$ corresponds to $S_\emptyset = 1$.

*Theorem* 3.13. The function $\tau \in \mathbb{C}[[p_1, p_2, \ldots]]$ is the exponential function of a solution of a KP-hierarchy if and only if, under the boson-fermion correspondence, its image can be represented by a decomposable wedge product:

$$\tau \leftrightarrow \varphi_1(z) \wedge \varphi_2(z) \wedge \ldots, \quad \varphi_i \in \mathbb{C}[z^{-1}][[z]]$$

*Geometric reformulation:* $\tau$-functions form a cone over the Grassmannian $Gr_{\frac{\infty}{2}}(V)$ of half-infinite subspaces Plücker embedded into the projective space $\mathbb{P}\Lambda^{\frac{\infty}{2}}(V)$.

Plücker embedding is defined by quadratic (Plücker) relations. These algebraic equations written in terms of coefficients of Taylor series of $\tau$ are called the *Hirota bilinear equations.*

KP-hierarchy (3) coincides with the collection of these equations written for $F = \log(\tau)$.

**Example 3.14.** $(z^{-1}+z^2)\wedge z^{-2}\wedge z^{-3}\wedge\cdots = v_\emptyset + v_3 \leftrightarrow S_\emptyset + S_3 = 1+\frac{1}{6}(p_1^3+3p_1p_2+2p_3)$. The function $\log(1+\frac{1}{6}(p_1^3+3p_1p_2+2p_3))$ is a solution of KP (3).

### 3.3. $\widehat{gl}(\infty)$

**Definition 3.15.** $gl(\infty)$ is the Lie algebra of DO on $S^1$.

Operator $A \in gl(\infty)$ acts on $V$ linearly. With respect to the basis $z^i$, $i \in \mathbb{Z}$ operator $A$ is written as an infinite matrix $(a_{ij})$: $A : z^j \mapsto \sum_i a_{ij} z^i$.

Lie algebra $gl(\infty)$ is graded by $\deg(z^i(\frac{\partial}{\partial z})^j) = i - j$. Operator $A$ induces the operator $\widehat{A} : \Lambda^{\frac{\infty}{2}} V \to \Lambda^{\frac{\infty}{2}} V$, $\hat{A}(z^{k_1}\wedge z^{k_2}\wedge\dots) = A(z^{k_1})\wedge z^{k_2}\wedge + z^{k_1}\wedge A(z^{k_2})\wedge\cdots+\dots$.

*Remark* 3.16. When $deg(A) = 0$, say $A = diag(\dots, a_{-1}, a_0, a_1, \dots)$, we can obtain divergence. Hence, we need *the normal ordering*, as follows. $\hat{A}(z^{k_1}\wedge z^{k_2}\wedge\dots) = \sum_{i=1}^{\infty}(a_{k_i} - a_{-i}) z^{k_1}\wedge z^{k_2}\wedge\dots$.

**Example 3.17.** $E = z\frac{\partial}{\partial z} \in gl(\infty)$, $E : z^k \mapsto kz^k$. Then $\hat{E}$ is the energy operator $\hat{E} : v_\lambda \mapsto |\lambda| v_\lambda$, where $|\lambda| = \sum \lambda_i$.

*Remark* 3.18. $A \mapsto \hat{A}$ is not a Lie algebra homomorphism.

**Example 3.19.** Abusing notations we denote by $z^m$ the operator of multiplication by $z^m$. Then $[z^m, z^n] = 0$, but $[\widehat{z^m}, \widehat{z^n}] = n\delta_{m,-n}$.

More exactly, $\widehat{[A,B]} = [\hat{A}, \hat{B}] +$ scalar operator. Therefore, we have a representation of $\widehat{gl}(\infty)$ - a one-dimensional central extension of $gl(\infty)$.

**Proposition 3.2.** *The Lie group action of $e^{t\hat{A}}$ preserves the set of decomposable vectors.*

*Proof.* $e^{t\hat{A}}(\varphi_1\wedge\varphi_2\wedge\dots) = e^{tA}\varphi_1\wedge e^{tA}\varphi_2\wedge\dots$ if diagonal entries of $A$ are all zeroes and $e^{t\hat{A}}(\varphi_1\wedge\varphi_2\wedge\dots) = e^{t(A-a_{-1})}\varphi_1\wedge e^{t(A-a_{-2})}\varphi_2\wedge\dots$ if $A = diag(\dots, a_{-1}, a_0, a_1, \dots)$. □

## References


[A] V.I. Arnold, *Mathematical methods of classical mechanics.* Graduate Texts in Mathematics, 60, Springer-Verlag, New York, 1989.

[K] M. Kazarian, *KP hierarchy for Hodge integrals*, arXiv:0809.3263

[MJD] T. Miwa, M. Jimbo, E. Date, *Solitons: Differential Equations, Symmetries and Infinite Dimensional Algebras*, Cambridge University Press, 2000.

[P] A.M. Perelomov, *Integrable systems of classical mechanics and Lie algebras. Vol. I*, Birkhäuser Verlag, Basel, 1990.



Department of Mathematics, Michigan State University, East Lansing, MI 48824-1027
*E-mail address*: mshapiro@math.msu.edu

Proceedings of 16$^{th}$ Gökova
Geometry-Topology Conference
pp. 34 – 57

# Integrable systems and holomorphic curves

*Paolo Rossi*

ABSTRACT. In this paper we attempt a self-contained approach to infinite dimensional Hamiltonian systems appearing from holomorphic curve counting in Gromov-Witten theory. It consists of two parts. The first one is basically a survey of Dubrovin's approach to bihamiltonian tau-symmetric systems and their relation with Frobenius manifolds. We will mainly focus on the dispersionless case, with just some hints on Dubrovin's reconstruction of the dispersive tail. The second part deals with the relation of such systems to rational Gromov-Witten and Symplectic Field Theory. We will use Symplectic Field theory of $S^1 \times M$ as a language for the Gromov-Witten theory of a closed symplectic manifold $M$. Such language is more natural from the integrable systems viewpoint. We will show how the integrable system arising from Symplectic Field Theory of $S^1 \times M$ coincides with the one associated to the Frobenius structure of the quantum cohomology of $M$.

## 1. A model for infinite-dimensional Hamiltonian systems

In this section we introduce the language and formalism necessary to study the kind of infinite dimensional Hamiltonian systems which are related to Gromov-Witten theory. We will define a formal (infinite dimensional) phase space with a Poisson structure and consider a space of functions over this phase space which, together with the Poisson structure, will give rise to a Hamiltonian dynamics described in the form of a system of PDEs with one spatial and one temporal variable. This material comes almost entirely from [8].

**Phase space and local functionals.** We will consider as our phase space a space of formal maps $\mathcal{L} = \mathcal{L}(\mathbb{R}^n) = \{u : S^1 \to \mathbb{R}^n\}$ from the circle to an $n$ dimensional real affine space. We will denote by $x$ the coordinate on $S^1$ and by $u^\alpha = u^\alpha(x)$ the $\alpha$-th component of the map $u$, with $\alpha = 1, \dots, n$. Formal here refers to the choice of the space of functions over $\mathcal{L}$ that we are going to make (indeed the maps $u$ and the space $\mathcal{L}$ itself should be seen as convenient notational tools to describe a theory which is completely defined at the level of the space of functions, see also Remark 1.1). As functions on $\mathcal{L}$ we will consider the space of local functionals of the form

$$F[u] = \frac{1}{2\pi} \int_0^{2\pi} f(x, u, u_x, u_{xx}, \dots) \mathrm{d}x$$

where $u_x$ denotes the derivative of $u$ with respect to $x$, $u_{xx}$ the second derivative, and so on and all of them are considered as independent formal variables. We will also denote by $u^{(k)}$ the $k$-th $x$-derivative of $u$ and by $u^{\alpha,k}(x)$ its $\alpha$-th component. The density $f(x, u, u_x, u_{xx}, \ldots)$ is a differential polynomial, i.e. a smooth function in the $x$ and $u$ variables (on $S^1 \times \mathbb{R}^n$) and a polynomial of any order in the higher derivatives. Usually we will denote a local functional and its density by higher and lower case versions of the same letter, respectively. Every time $f(x, u, u_x, u_{xx}, \ldots)$ coincides with a total $x$-derivative via the chain rule

$$\partial_x g(x, u, u_x, u_{xx}, \ldots) := \frac{\partial g}{\partial x} + \sum_{\alpha,k} \frac{\partial g}{\partial u^{\alpha,k}} u^{\alpha,k+1}$$

then

$$F[u] = \frac{1}{2\pi} \int_0^{2\pi} f(x, u, u_x, u_{xx}, \ldots) \mathrm{d}x = \frac{1}{2\pi} \int_0^{2\pi} \partial_x g(x, u, u_x, u_{xx}, \ldots) \mathrm{d}x = 0$$

This is of course equivalent to considering the space of local functionals as the space of differential polynomials $f(x, u, u_x, u_{xx}, \ldots)$ up to total $x$-derivatives.

**Poisson brackets.** We want to endow our formal loop space $\mathcal{L}$ with a Poisson structure. The Poisson brackets between two local functionals will have the form

$$\{F[u], G[u]\} = \frac{1}{4\pi^2} \int_0^{2\pi} \int_0^{2\pi} \frac{\delta F}{\delta u^\alpha(x)} \{u^\alpha(x), u^\beta(y)\} \frac{\delta G}{\delta u^\beta(y)} \,\mathrm{d}x\,\mathrm{d}y$$

where

$$\frac{\delta F}{\delta u^\alpha(x)} := \sum_s (-1)^s \partial_x^s \frac{\partial f}{\partial u^{\alpha,s}}$$

$$\{u^\alpha(x), u^\beta(y)\} := \sum_{s=0}^{\infty} a_s^{\alpha\beta}(u, u_x, u_x x, \ldots) \delta^{(s)}(x - y)$$

and where $\delta^{(s)}(x - y) = \partial_x^s \delta(x - y)$ is a formal delta-distribution defined by

$$\frac{1}{2\pi} \int_0^{2\pi} f(x, u, u_x, u_{xx}, \ldots) \delta^{(s)}(x - y) \,\mathrm{d}x := \partial_x^s f(x, u, u_x, u_{xx}, \ldots)|_{x=y}$$

We then see that, after the action of the delta-function (which elides one of the integrals), the Poisson bracket of two local functionals is still a local functional. Of course, in order to have a genuine Poisson bracket, the usual antisymmetry, Leibnitz and Jacobi identities must hold, which imposes some constraints on the form of the coefficients $a_s^{\alpha\beta}(u, u_x, u_{xx}, \ldots)$. We will study below some of such constraints.

**Remark 1.1.** Notice that a possible choice of coordinates on the space $\mathcal{L}$ is given by the formal Fourier components of the formal functions $u^\alpha(x)$ on the circle. The functions $u^\alpha(x)$ and their Fourier components are formal in the sense that it is coherent with our choice of formal local functionals that we don't require any sort of convergence for the Fourier series of $u^\alpha(x)$. More precisely, besides the infinite set of formal variables

$u, u_x, u_{xx}, \ldots$, we consider another infinite set of formal variables $\{u_0^\alpha, p_k^\alpha, q_k^\alpha\}$ which we want to see as Fourier coefficients in a formal Fourier series:

$$u^\alpha(x) = u_0^\alpha + \sum_{k=1}^{\infty} \left(p_k^\alpha \mathrm{e}^{ikx} + q_k^\alpha \mathrm{e}^{-ikx}\right)$$

The above formula and its formal $x$-derivatives (term by term in the Fourier series) should be seen as a change of coordinates between $u, u_x, u_{xx}, \ldots$ and the $\{u_0^\alpha, p_k^\alpha, q_k^\alpha\}$ variables. A local functional $F[u]$ can then be written as an infinite polynomial in the $\{u_0^\alpha, p_k^\alpha, q_k^\alpha\}$ variables (i.e. a finite degree infinite sum of monomials in $\{u_0^\alpha, p_k^\alpha, q_k^\alpha\}$) just by substituting $u, u_x, u_{xx}, \ldots$ with the corresponding Fourier seris and then formally integrating $f(x, u, u_x, u_{xx}, \ldots)$ term by term in $x$. The definition (or inverse transformation) of $u_0^\alpha$, $p_k^\alpha$ and $q_k^\alpha$ in terms of the formal map $u$ is given as local functionals by

$$u_0^\alpha = \frac{1}{2\pi} \int_0^{2\pi} u^\alpha \, \mathrm{d}x$$

$$p_k^\alpha = \frac{1}{2\pi} \int_0^{2\pi} u^\alpha \mathrm{e}^{-ikx} \, \mathrm{d}x$$

$$q_k^\alpha = \frac{1}{2\pi} \int_0^{2\pi} u^\alpha \mathrm{e}^{+ikx} \, \mathrm{d}x$$

Coherently the delta-function has the Fourier series

$$\delta(x - y) = \sum_{k=-\infty}^{+\infty} \mathrm{e}^{ik(x-y)}$$

from which one can compute the Poisson bracket $\left\{p_k^\alpha, q_l^\beta\right\}$.

**Vector fields and Hamiltonian vector fields.** The vector fields on our formal loop space will have the general form

$$a = a^0(x, u, u_x, u_{xx}, \ldots)\frac{\partial}{\partial x} + \sum_{\beta,k} a^{\beta,k}(x, u, u_x, u_{xx}, \ldots)\frac{\partial}{\partial u^{\beta,k}}$$

and their action as differential operators on local functionals is of course

$$a(F) := \frac{1}{2\pi} \int_0^{2\pi} a(f) \mathrm{d}x$$

We will be interested in particular in *evolutionary vector fields*, which are characterized by the vanishing of the $\frac{\partial}{\partial x}$ component and the property $[\partial_x, a] = 0$. Notice that, in particular, $\partial_x = \frac{\partial}{\partial x} + u^{\alpha,k+1}\frac{\partial}{\partial u^{\alpha,k}}$ is an evolutionary vector field. It is straightforward to check that from the definition of evolutionary vector field it follows that its general form is

$$a = \sum_{\beta,k} \partial_x^k a^\beta \frac{\partial}{\partial u^{\beta,k}}$$

for some differential polynomials $a^1(x, u, u_x, u_{xx}, \ldots), \ldots, a^n(x, u, u_x, u_{xx}, \ldots)$. In case such differential polynomials do not depend explicitly on $x$, the evolutionary vector field will be called *translation invariant.* As usual the presence of a Poisson bracket and a selected Hamiltonian function $H$ (functional $H[u]$, in our case) generates a dynamic on the phase space in the form of a Hamiltonian vector field. On the formal loop space $\mathcal{L}$ the Hamiltonian flow of $H[u]$ has infinitesimal expression (formal Hamiltonian vector field, or Hamilton's equations)

$$u_t^\alpha = \{u^\alpha(x), H[u]\} = \frac{1}{2\pi}\int_0^{2\pi} \{u^\alpha(x), u^\beta(y)\} \frac{\delta H}{\delta u^\beta(y)}\,\mathrm{d}y = U^\alpha(u, u_x, u_{xx}, \ldots).$$

More precisely the above components $U^\alpha(u, u_x, u_{xx}, \ldots)$ are the $n$ differential polynomials defining the *evolutionary* Hamiltonian vector field of $H[u]$. This, for $\alpha = 1, \ldots, n$, can be seen as a system of evolutionary PDEs for the functions $u^\alpha = u^\alpha(x,t)$ representing the Hamiltonian flow on $\mathcal{L}$. We will call a system of this type an infinte-dimensional Hamiltonian system. When the Hamiltonian does not depend on $x$ explicitly, its Hamiltonian vector field is translation invariant.

**Example 1.2.** The most celebrated example of infinite-dimensional Hamiltonian system (because of all its beautiful properties and its appearence in many different contexts of mathematics) is Korteweg-deVries (KdV for short) equation. In our language it corresponds to the choice of the formal loop space $\mathcal{L} = \mathcal{L}(\mathbb{R})$ (so $u$ has only one component $u(x)$) and the Poisson structure $\{u(x), u(y)\} = \delta'(x-y)$. The Hamiltonian functional generating the KdV flow is $H[u] = \frac{1}{2\pi}\int_0^{2\pi}\left(\frac{u^3}{6} + \frac{1}{24}(u_x^2 + 2u_{xx})\right)dx$. Hence the KdV equation (Hamilton's equation for $H[u]$) is

$$u_t = uu_x + \frac{1}{12}u_{xxx}$$

**Long wave limit and extended formal loop space.** In order to have an extra formal variable keeping track of the number of $x$-derivatives entering our formulas it is costumary to rescale both the $x$ and $t$ variables by

$$x \mapsto \frac{x}{\epsilon} \qquad\qquad t \mapsto \frac{t}{\epsilon}$$

The limit $\epsilon \to 0$ is called the long wave limit, in reference to its effect on the behavior of solutions to the Hamiltonian PDEs. Notice that the definition of the delta-function implies the corresponding rescaling $\delta(x,y) \to \frac{1}{\epsilon}\delta(x-y)$. The general local functional hence becomes a polynomial in $\epsilon$. However we will, from now on, extend our class of local functionals including *formal power series* in $\epsilon$ of the form

$$F[u] = \frac{1}{2\pi}\int_0^{2\pi}\left(\sum_{k=0}^{\infty} f_k(x,u,\dots,u^{(k)})\epsilon^k\right)\mathrm{d}x$$

where, assigning degree $i$ to the variable $u^{\alpha,i}$, $f_k$ has degree $k$.

The general Poisson bracket takes the form

$$\begin{aligned}\{u^\alpha(x),u^\beta(y)\} =& \frac{1}{\epsilon}h^{\alpha\beta}(u)\delta(x-y)\\ &+ g^{\alpha\beta}(u)\delta'(x-y)+\Gamma^{\alpha\beta}_\gamma(u)u^\gamma_x\delta(x-y)\\ &+O(\epsilon)\end{aligned} \tag{1}$$

**Poisson brackets of hydrodynamic type.** Poisson brackets for which the $\frac{1}{\epsilon}$-term vanishes are called *hydrodynamic.* They were studied by Dubrovin and Novikov and their main property is

**Theorem 1.1** ([7]). *Let*

$$\{u^\alpha(x),u^\beta(y)\} = g^{\alpha\beta}(u)\delta'(x-y)+\Gamma^{\alpha\beta}_\gamma(u)u^\gamma_x\delta(x-y)+O(\epsilon)$$

*be a Poisson bracket of hydrodynamic type. Then (from the antisymmetry, Jacobi and Liebniz properties):*

(i) $g^{\alpha\beta}$ *is symmetric*
(ii) *if* $g^{\alpha\beta}$ *is nondegenerate, then it is a flat metric and* $\Gamma^{\alpha\beta}_\gamma(u) = -g^{\alpha\mu}\Gamma^\beta_{\mu\gamma}(u)$ *are its Christoffel symbols.*

We will assume from now on that the matrix $g^{\alpha\beta}$ is indeed nondegenerate. This result implies in particular that, if we choose flat coordinates for such a metric, $g^{\alpha\beta}(u)$ is mapped to a constant matrix $\eta^{\alpha\beta}$ and the term involving Christoffels vanishes.

In fact it turns out that the full tail of positive order in $\epsilon$ in the Poisson bracket can be also killed if we allow more general changes of our formal variables in the form of a *Miura transformation*

$$u^\alpha \mapsto \tilde{u}^\alpha := \sum_{k=0}^{\infty}\tilde{u}^\alpha_k(u,u_x,,\dots,u^{(k)})\epsilon^k$$

where $\det(\frac{\partial\tilde{u}^\alpha_0(u)}{\partial u^\beta})\neq 0$ and assigning degree $i$ to the variable $u^{\alpha,i}$, $\tilde{u}^\alpha_k(u,u_x,\dots,u^{(k)})$ is a polynomial of degree $k$. Notice that such transformations indeed form a group, where inversion consists in solving an infinite order ODE for $u^\alpha$ recursively (and completely algebraically) in power series of $\epsilon$ and in terms of the known functions $\tilde{u}^\beta$ and their $x$-derivatives. This in fact is our main motivation for having introduced the long wave limit variable $\epsilon$ and the extended formal loop space. We have in particular

**Theorem 1.2** ([8]). *Given a Poisson bracket of hydrodynamic type, there exist a Miura transformation*

$$u^\alpha \mapsto \tilde{u}^\alpha := \sum_{k=0}^{\infty} \tilde{u}_k^\alpha(u, u_x, , \ldots, u^{(k)})\epsilon^k$$

*such that, in the new coordinates, the Poisson bracket is reduced to the normal form*

$$\{u^\alpha(x), u^\beta(y)\} = \eta^{\alpha\beta}\delta'(x-y)$$

*The local functionals* $\frac{1}{2\pi}\int_0^{2\pi} \tilde{u}^\alpha(u, u_x, u_{xx}, \ldots; \epsilon)\, \mathrm{d}x$ *are Casimirs of the Poisson bracket (i.e. they commute with any other local functional). In particular* $\tilde{u}_0^\alpha(u)$ *are flat coordinates for the metric* $g^{\alpha\beta}(u)$.

Denoting by $\{\tilde{u}_0^\alpha, \tilde{p}_k^\alpha, \tilde{q}_k^\alpha\}$ the formal Fourier components of $\tilde{u}^\alpha$ we see that their only nonzero Poisson bracket is

$$\{\tilde{p}_k^\alpha, \tilde{q}_l^\beta\} = \mathrm{i}\, k\, \eta^{\alpha\beta}\, \delta_{k,l}$$

**Example 1.3.** There is another Poisson structure and Hamiltonian giving rise to the KdV equation on $\mathcal{L}(\mathbb{R})$. Namely,

$$\{u(x), u(y)\} = u\delta'(x-y) + \frac{1}{2}u_x\delta(x-y) + \frac{\epsilon^2}{8}\delta'''(x-y)$$

$$H[u] = \frac{1}{2\pi}\int_0^{2\pi} \frac{u^2}{2}\mathrm{d}x$$

The Miura transformation reducing such Poisson structure to normal form is found by solving recursively in $\epsilon$ the Riccati equation

$$u = \frac{\tilde{u}^2}{4} - \frac{\epsilon}{2\sqrt{2}}\tilde{u}_x$$

**Bihamiltonian structures of hydrodynamic type.** Following Magri [15], we say that two Poisson structures of hydrodynamic type $\{\cdot,\cdot\}_1$ and $\{\cdot,\cdot\}_2$ form a *bihamiltonian structure* if and only if the linear combination $\{\cdot,\cdot\}_\lambda := \{\cdot,\cdot\}_2 - \lambda\{\cdot,\cdot\}_1$ is still a Poisson structure of hydrodynamic type. The following theorem gives a way to find commuting Hamiltonians for a bihamiltonian structure.

**Theorem 1.3** ([15]). *Let* $\{u^\alpha(x), u^\beta(y)\}_\lambda$ *be reduced to the normal form* $-\lambda\eta^{\alpha,\beta}\delta'(x-y)$ *by the Miura transformation*

$$u^\alpha \mapsto \tilde{u}^\alpha(u, u_x, u_{xx}, \ldots; \epsilon, \lambda) = \sum_{p=-1}^{\infty} \frac{1}{\lambda^{p+1}} f_p^\alpha(u, u_x, u_{xx}, \ldots; \epsilon).$$

*Then*

$$\{\cdot, F_{p+1}^\alpha\}_1 = \{\cdot, F_p^\alpha\}_2$$

*and, in particular,*

$$\{F_p^\alpha, F_q^\beta\}_1 = \{F_p^\alpha, F_q^\beta\}_2 = 0$$

*where* $F_p^\alpha[u] = \frac{1}{2\pi}\int_0^{2\pi} f_p^\alpha(u, u_x, u_{xx}, \ldots; \epsilon)\mathrm{d}x$.

By definition, the new variables $\tilde{u}^\alpha$ are densities of Casimirs of the Poisson bracket $\{u^\alpha(x), u^\beta(y)\}_\lambda$.

**Example 1.4.** The KdV Poisson structure

$$\{u(x), u(y)\}_1 = \delta'(x-y)$$

$$\{u(x), u(y)\}_2 = u\delta'(x-y) + \frac{1}{2}u_x\delta(x-y) + \frac{\epsilon^2}{8}\delta'''(x-y)$$

form a bihamiltonian structure. $\{u(x), u(y)\}$ is reduced to normal form by solving the Riccati equation

$$\frac{\mathrm{i}\epsilon}{2\sqrt{2\lambda}}\tilde{u}_x - \frac{\tilde{u}^2}{4\lambda} = u - \lambda$$

recursively in $\epsilon$ and $\lambda$ as

$$\tilde{u} = \sum_{m=0}^{\infty} \frac{\tilde{u}_m}{(\sqrt{\lambda})^m} + 2\lambda.$$

We find in this way that $\tilde{u}_m$ is a total $x$-derivative when $m$ is odd and we define

$$F_p = \frac{1}{2\pi}\int_0^{2\pi} \tilde{u}_{2p+2}\mathrm{d}x.$$

Such Hamiltonians are symmetries of the KdV equation.

Notice that, although $F_p^\alpha$ form an infinite number of commuting Hamiltonians, they may very well fail to be linearly independent. The following example is instructive.

**Example 1.5.** On the loop space $\mathcal{L}(\mathbb{R}^2)$ with coordinates $u(x), v(x)$ consider the bihamiltonian structure

$$\{u(x), u(y)\}_1 = \{v(x), v(y)\}_1 = 0$$

$$\{u(x), v(y)\}_1 = \frac{1}{\epsilon}(\delta(x-y+\epsilon) - \delta(x-y))$$

and

$$\{u(x), u(y)\}_2 = \frac{1}{\epsilon}(e^{v(x+\epsilon)}\delta(x-y+\epsilon) - e^{v(x)}\delta(x-y+\epsilon))$$

$$\{v(x), v(y)\}_2 = \frac{1}{\epsilon}(\delta(x-y+\epsilon) - \delta(x-y-\epsilon))$$

$$\{u(x), v(y)\}_2 = \frac{1}{\epsilon}u(x)(\delta(x-y+\epsilon) - \delta(x-y))$$

Applying the first one to the Hamiltonian

$$H[u,v] = \frac{1}{2\pi}\int_0^{2\pi}\left(\frac{1}{2}u^2 + e^v\right)\mathrm{d}x$$

one obtains the Toda equations

$$u_t = \frac{1}{\epsilon}(e^{v(x+\epsilon)} - e^{v(x)}) = (e^v)_x + \frac{\epsilon}{2}(e^v)_{xx} + \dots$$

$$v_t = \frac{1}{\epsilon}(u(x) - u(x-\epsilon)) = v_x - \frac{\epsilon}{2}v_{xx} + \dots$$

We can apply the above theorem to find mutually commuting Hamiltonians for the Toda bihamiltonian structure (incidentally containing the Toda Hamiltonian $H[u,v]$ itself). Let us study the Casimirs for the first Poisson bracket. Its reducing transformation is given by

$$v = \tilde{v}$$

$$u = \frac{1}{\epsilon}(\tilde{u}(x+\epsilon) - \tilde{u}(x))$$

whose inverse is

$$\tilde{v} = v$$

$$\tilde{u} = \epsilon\left(\frac{\partial_x}{e^{\epsilon\partial_x} - 1}\right)u = \sum_{n=0}^{\infty}\frac{B_n}{n!}(\epsilon\partial_x)^n u$$

where $B_n$ are the Bernoulli numbers. Hence, the Casimirs of $\{\cdot,\cdot\}_1$ are $\frac{1}{2\pi}\int_0^{2\pi} v\,\mathrm{d}x$ and $\frac{1}{2\pi}\int_0^{2\pi}\tilde{u}\,\mathrm{d}x = \frac{1}{2\pi}\int_0^{2\pi} u\,\mathrm{d}x$.

Notice now that $\frac{1}{2\pi}\int_0^{2\pi} v\,\mathrm{d}x$ is a Casimir of $\{\cdot,\cdot\}_2$ too! This implies that the one component of the reducing transformation for $\{\cdot,\cdot\}_\lambda = \{\cdot,\cdot\}_2 - \lambda\{\cdot,\cdot\}_1$ is trivial

$$\tilde{v}(u, v, u_x, v_x, \dots; \epsilon, \lambda) = v$$

so that one of the two infinite sequences of Hamiltonians $F_p^\alpha[u,v]$ vanishes.

**Regular Poisson pencils.** We plan now to give a definition of a bihamiltonian system which is sufficiently complete, meaning that its symmetries span a sufficiently large space of local functionals. To this end, starting from a bihamiltonian structure of hydrodynamic type on the loop space $\mathcal{L}(\mathbb{R}^n)$ and its associated bihamiltonian system $F_p^\alpha[u]$, we define its *space of conserved quantities* as

$$\mathcal{I} = \left\{h(u, u_x, u_{xx}, \dots; \epsilon) \mid \{H[u], F_p^\alpha[u]\}_1 = 0\right\}/\mathbb{C}$$

$$\tilde{\mathcal{I}} = \mathcal{I}/\mathrm{Im}\partial_x = \left\{H[u] = \frac{1}{2\pi}\int_0^{2\pi} h(u, u_x, u_{xx}, \dots; \epsilon)\,\mathrm{d}x \mid h \in \mathcal{I}\right\}$$

These are in general infinite dimensional spaces so, in order to control their size, we will use a filtration which is compatible with the bihamiltonian structure. Namely a *compatible flag of conserved quantities* for the bihamiltonian system $F_p^\alpha[u]$ is a flag

$$\mathcal{F}_{-1}\mathcal{I} \subset \mathcal{F}_0\mathcal{I} \subset \ldots \subset \mathcal{I}$$

inducing a flag

$$\mathcal{F}_{-1}\tilde{\mathcal{I}} \subset \mathcal{F}_0\tilde{\mathcal{I}} \subset \ldots \subset \tilde{\mathcal{I}}$$

such that

$$\dim \mathcal{F}_{-1}\mathcal{I} = \dim \mathcal{F}_{-1}\tilde{\mathcal{I}} = \dim \frac{\mathcal{F}_k\mathcal{I}}{\mathcal{F}_{k-1}\mathcal{I}} = \dim \frac{\mathcal{F}_k\tilde{\mathcal{I}}}{\mathcal{F}_{k-1}\tilde{\mathcal{I}}} = n \qquad k \geq 0$$

and such that

$$\begin{aligned} \{\cdot, \mathcal{F}_{-1}\tilde{\mathcal{I}}\} &= 0 \\ \{\cdot, \mathcal{F}_{k-1}\tilde{\mathcal{I}}\}_2 &\subset \{\cdot, \mathcal{F}_k\tilde{\mathcal{I}}\}_1 \qquad k \geq 0 \\ \{\mathcal{F}_i\tilde{\mathcal{I}}, \mathcal{F}_j\tilde{\mathcal{I}}\}_1 = 0 & \qquad i, j \geq -1. \end{aligned}$$

A bihamiltonian structure of hydrodynamic type possessing a compatible flag of conserved quantities is called *regular*. A bihamiltonian structure of hydrodynamic type whose associated system $F_p^\alpha[u]$ spans a compatible flag is called *nonresonant*. For instance, as we saw, the bihamiltonian structure of Example 1.5 is resonant. However we will see how it is regular anyway.

**Tau-symmetry.** A regular bihamiltonian structure is said to possess a *$\tau$-symmetry* (or to be *$\tau$-symmetric*) if and only if there exist a basis of the corresponding flag of conserved quantities

$$\begin{aligned} \langle h_{1,-1}, \ldots, h_{n,-1}\rangle &= \mathcal{F}_{-1}\mathcal{I} \\ \langle h_{1,p}, \ldots, h_{n,p}\rangle &= \frac{\mathcal{F}_p\mathcal{I}}{\mathcal{F}_{p-1}\mathcal{I}} \qquad p \geq 0 \end{aligned}$$

such that

$$\{h_{\alpha,p-1}, H_{\beta,q}\}_1 = \{h_{\beta,q-1}, H_{\alpha,p}\}_1 =: \partial_x \Omega_{\alpha,p;\beta,q}.$$

Here, as usual, the Poisson bracket of a local functional $H_{\beta,q}[u]$ and a density $h_{\alpha,p_1}(u, u_x, u_{xx}, \ldots; \epsilon)$ is a differential polynomial. In our specific case such polynomial has to be a total $x$-derivative, since $H_{\beta,p}[u]$ and $H_{\alpha,p}[u]$ Poisson-commute for any $\alpha, p, \beta, q$. Hence the definition of the symmetric matrix $\Omega_{\alpha,p;\beta,q}(u, u_x, u_{xx}, \ldots; \epsilon)$.

We will include in our definition of $\tau$-symmetric basis the requirement that $h_{1,0}(u, u_x, u_{xx}, \ldots; \epsilon)$ is the generator of spacial translations, i.e.

$$\{u^\alpha, H_{1,0}\}_1 = u^\alpha_x.$$

This implies that any solution to the associated Hamiltonian system of PDEs will be a function $u^\alpha = u^\alpha(x + t^{1,0}, t; \epsilon)$ so that we can indeed identify the $x$ and $t^{1,0}$ variables, using the notation $u^\alpha = u^\alpha(t; \epsilon)$.

From now on we will make the Miura transformation to normal coordinates for the first Poisson bracket

$$u^\alpha \mapsto \eta^{\alpha\beta} h_{\beta,-1}(u, u_x, u_{xx}, \ldots; \epsilon).$$

For simplicity we will still denote by $u^\alpha$ such normal coordinates.

The dynamical system generated by the densities $h_{\alpha,p}$ is called a *bihamiltonian $\tau$-symmetric (BHTS) hierarchy.*

The main property of a BHTS hierarchy is given by the following theorem.

**Theorem 1.4** ([8]). *For any solution $u^\alpha = u^\alpha(t;\epsilon)$ of a BHTS hierarchy, there exists a function $\tau = \tau(t;\epsilon)$ such that*

$$\Omega_{\alpha,p;\beta,q}(u(t;\epsilon), u_x(t;\epsilon), \ldots; \epsilon) = \epsilon^2 \frac{\partial \log \tau}{\partial t^{\alpha,p} \partial t^{\beta,q}}.$$

*In particular, in normal coordinates,*

$$u^\alpha(t;\epsilon) = \epsilon^2 \frac{\partial \log \tau}{\partial x \partial t^{\alpha,0}}.$$

**Frobenius manifolds.** An important result in Dubrovin's work (see e.g. [8]) is the classification of the $\epsilon = 0$ limit of BHTS hierarchies in terms of Frobenius manifolds. The abstract definition of a Frobenius manifold can be found easily in the literature ([6], [8]), but we prefer to stick to a more operational definition, as in [5]. A *Frobenius manifold* structure on an open subset $U \subset \mathbb{R}^n$ is a (smooth or analytic) solution $F : U \to \mathbb{R}$ of the *Witten-Dijkgraaf-Verlinde-Verlinde (WDVV)* equations

$$\frac{\partial^3 F}{\partial u^\alpha \partial u^\beta \partial u^\gamma} \eta^{\gamma\delta} \frac{\partial^3 F}{\partial u^\delta \partial u^\epsilon \partial u^\mu} = \frac{\partial^3 F}{\partial u^\mu \partial u^\beta \partial u^\gamma} \eta^{\gamma\delta} \frac{\partial^3 F}{\partial u^\delta \partial u^\epsilon \partial u^\alpha}$$

such that $\eta^{\gamma\delta}$ is the inverse matrix of the *nondegenerate constant* metric

$$\eta_{\gamma\delta} := \frac{\partial^3 F}{\partial u^1 \partial u^\gamma \partial u^\delta} = \text{constant}$$

and such that $F = F(u^1, \ldots, u^n)$ is *quasi-homogeneous*, i.e. there exist a linear *Euler vector field*

$$E = \sum_{\alpha=1}^{n} d_\alpha u^\alpha \frac{\partial}{\partial u^\alpha} + \sum_{\alpha | d_\alpha \neq 0} r^\alpha \frac{\partial}{\partial u^\alpha}$$

on $U$ for which

$$E(F) = E^\alpha \frac{\partial F}{\partial u^\alpha} = d_F F + A_{\alpha\beta} u^\alpha u^\beta + B_\alpha u^\alpha + C$$

where $d_\alpha, r^\alpha, d_F, A_{\alpha\beta}, B_\alpha, C$ are constants.

**Remark 1.6.** Notice that any linear vector field $E$ of the form $E = (a^\alpha_\beta + r^\alpha)\frac{\partial}{\partial u^\alpha}$ with $a^\alpha_\beta$ a constant diagonalizable matrix can always be reduced, by a linear change of variables, to the form

$$E = \sum_{\alpha=1}^{n} d_\alpha u^\alpha \frac{\partial}{\partial u^\alpha} + \sum_{\alpha | d_\alpha \neq 0} r^\alpha \frac{\partial}{\partial u^\alpha}.$$

Usually one distinguishes between the case $d_1 \neq 0$, for which the normalization $d_1 = 1$ is chosen, and the case $d_1 = 0$, for which the Frobenius manifold is called *degenerate*.

Let

$$c_{\alpha\beta\gamma} = \frac{\partial^3 F}{\partial u^\alpha \partial u^\beta \partial u^\gamma}.$$

We will always use the flat metric $\eta$ to raise and lower the indices of tensor fields on $U$. In particular we obtain a Frobenius algebra structure on the tangent bundle of $U$ whose structure constants are $c^\alpha_{\beta\gamma} = \eta^{\alpha\mu}\frac{\partial^3 F}{\partial u^\mu \partial u^\beta \partial u^\gamma}$. This algebra is associative thanks to WDVV equations and $e := \frac{\partial}{\partial u^1}$ is the identity. We will denote by $\star$ the product in such Frobenius algebra. Notice that $\eta_{\alpha\beta} = c_{1\alpha\beta}$.

Two Frobenius structures on $U$ and $U'$ are *equivalent* if there exists a diffeomorphism (or an analytic transformation) $\phi : U \to U'$ such that $\phi$ is a linear conformal transformation of the metrics $\eta$ and $\eta'$

$$\phi^* \eta' = c^2 \eta, \qquad c \in \mathbb{R}^*$$

and it induces an isomorphism of Frobenius algebras

$$\phi_* : (T_u U, \star) \to (T_{\phi(u)} U', \star').$$

Notice that this does not necessarily imply $F = \phi^* F'$.

**From Frobenius manifolds to dispersionless BHTS hierarchies.** From a Frobenius manifold one can recover a dispersionless (i.e. $\epsilon = 0$) BHTS hierarchy on $\mathcal{L}(\mathbb{R}^n)$ in the following way. Define a new metric (on the cotangent bundle of $U$) as $g_{\alpha\beta} := E_\gamma c^\gamma_{\alpha\beta}$. The metrics $\eta$ and $g$ define a bihamiltonian structure of hydrodynamic type as

$$\begin{aligned} \{u^\alpha(x), u^\beta(y)\}_1 &= \eta^{\alpha\beta}\delta'(x-y) \\ \{u^\alpha(x), u^\beta(y)\}_2 &= g^{\alpha\beta}(u)\delta'(x-y) + \Gamma^{\alpha\beta}_\gamma(u) u^\gamma_x \delta(x-y) \end{aligned}$$

We define, after Dubrovin, a *deformed flat connection* on $U \times \mathbb{C}^*$ given by

$$\tilde{\nabla}_X Y = X^\alpha \frac{\partial}{\partial u^\alpha} Y + \zeta X \star Y$$

$$\tilde{\nabla}_{\frac{\partial}{\partial \zeta}} Y = \frac{\partial}{\partial \zeta} Y - E \star Y - \frac{1}{\zeta}\mathcal{V} Y$$

where $X(\zeta), Y(\zeta) \in \Gamma(TU)$, $\zeta \in \mathbb{C}^*$ and $\mathcal{V}^\alpha_\beta = \frac{2-n}{2}\delta^\alpha_\beta + \frac{\partial E^\alpha}{\partial u^\beta}$. Dubrovin showed (see in particular [6], Lecture 2) that a fundamental solution exists for such connection in the form

$$Y^\alpha_\beta = \eta^{\alpha\mu}\frac{\partial h_\mu(u,\zeta)}{\partial u^\beta}\zeta^{\mathcal{V}}\zeta^R$$

where $R$ is a constant matrix (part of the so called *monodromy data* of the Frobenius manifold, whose explicit form can be recovered explicitly by studying the monodromy at $z=0$ of the the second flatness equation $\partial_z Y = E \star Y + \zeta^{-1}\mathcal{V}Y$) and

$$h_\alpha(u,\zeta) = \sum_{k=0}^{\infty} h_{\alpha,k-1}(u)\zeta^k.$$

The functions $h_{\alpha,p}$, for $p=-1,0,1,\ldots$, then form a BHTS hierarchy with respect to the above bihamiltonian structure. In particular we get

$$h_{\alpha,-1} = \eta_{\alpha\beta}u^\beta$$
$$h_{\alpha,0} = \frac{\partial F}{\partial u^\alpha}$$
$$\Omega_{\alpha,0;\beta,0} = \frac{\partial^2 F}{\partial u^\alpha \partial u^\beta}.$$

One can in fact show (see [8]) that any dispersionless BHTS hierarchy comes from a Frobenius manifold via this construction (up to a minor technical assumption about semisimplicity of the Frobenius algebras), yealding a full classification of the $\epsilon = 0$ limit.

Frobenius manifolds also provide the link between integrable systems and Gromov-Witten theory, since there is a canonical Frobenius structure on the quantum cohomology of a target symplectic manifold $M$ (assume for simplicity that $H^i(M)=0$ for $i$ odd, in order to avoid grading issues), its Frobenius potential being given by the genus 0 Gromov-Witten potential without descendants of $M$. In the case of quantum cohomology the Euler vector field is easily determined by the grading conditions on the Gromov-Witten potential, the matrix $R$ is given by the matrix of cup product with the Chern class of $M$, the metric $\eta$ is given by Poincaré pairing (the cohomology variables withouth gravitational descendants $t^{\alpha,0}$ playing the role of flat coordinates $u^\alpha$) and the fundamental solution to the flatness equations is provided by

$$h_{\alpha,p} = \left.\frac{\partial \mathbf{f}(t)}{\partial t^{\alpha,p}}\right|_{t^{\alpha,p}=0,\, p>0}$$

where $\mathbf{f}(t)$ is the rational Gromov-Witten potential of $M$ with descendants and $t$ globally refers to the cohomology varaibles with descendants $t^{\alpha,p}$, $p=0,1,\ldots$.

We will see this in detail in the next sections, using the language of Symplectic Field Theory of $M \times S^1$, which will lead us directly to the hierarchy $h_{\alpha,p}$ without passing explicitly through the Frobenius structure.

**Example 1.7.** The only one-dimensional Frobenius manifold (up to equivalence) is given by the potential
$$F(u) = \frac{u^3}{6}.$$
It corresponds to the quantum cohomology of a point and gives rise to the dispersionless KdV (also called Riemann or Burgers) hierarchy
$$\{u(x), u(y)\}_1 = \delta'(x-y)$$
$$\{u(x), u(y)\}_2 = u\delta'(x-y) + \frac{1}{2}u_x\delta(x-y)$$
and
$$\mathcal{V} = R = \begin{pmatrix} 0 & 0 \\ 0 & 0 \end{pmatrix}$$
$$h_p = \frac{u^{p+2}}{(p+2)!}.$$

**Example 1.8.** The Frobenius manifold given by the potential
$$F(u,v) = \frac{u^2 v}{2} + e^v$$
corresponds to the quantum cohomology of $\mathbb{P}^1$ and gives rise to the dispersionless Toda hierarchy
$$\{u(x), u(y)\}_1 = \{v(x), v(y)\}_1 = 0$$
$$\{u(x), v(y)\}_1 = \delta'(x-y)$$
$$\{u(x), u(y)\}_2 = e^{v(x)}v_x\delta(x-y)$$
$$\{v(x), v(y)\}_2 = 2\delta'(x-y)$$
$$\{u(x), v(y)\}_2 = u(x)\delta'(x-y)$$
and
$$\mathcal{V} = \begin{pmatrix} -\frac{1}{2} & 0 \\ 0 & \frac{1}{2} \end{pmatrix}$$
$$R = \begin{pmatrix} 0 & 0 \\ 2 & 0 \end{pmatrix}$$
$$h_1(u,v,\zeta) = -2e^{zu}\sum_{m\geq 0}\left(\gamma - \frac{1}{2}v + \digamma(m+1)\right)e^{mv}\frac{z^{2m}}{(m!)^2}$$
$$h_2(u,v,\zeta) = z^{-1}\sum_{m\geq 0}\left(e^{mv+zu}\frac{z^{2m}}{(m!)^2} - 1\right)$$
where $\gamma$ is Euler's constant and $\digamma$ is the digamma function.

Notice how the structure of the Frobenius manifold provided us with two infinite sequences of commuting Hamiltonians, whereas the technique of Theorem 1.3 only provided one (equivalent, up to a triangular transformation, to $h_{1,p}$). This reflects the regular but resonant nature of Toda hierarchy.

**Solving dispersionless BHTS hierarchies: the hodograph method.** What justifies the name of *integrable system* for a BHTS hierarchy is the possibility of finding a sufficiently complete set of solutions to the equations in a sufficiently explicit form. For dispersionless hierarchies this is performed via the generalized hodograph method ([17],[8]), for which we now give a simple but clear proof. Let us start by remarking a very important fact. The Hamiltonians $H_{\alpha,p}[u] = \frac{1}{2\pi}\int_0^{2\pi} h_{\alpha,p}(u)\mathrm{d}x$ form a complete family of symmetries in the following sense.

**Theorem 1.5** ([8]). *For any symmetry $\frac{1}{2\pi}\int_0^{2\pi} f(u^1,\ldots,u^n)\mathrm{d}x$ of a dispersionless BHTS hierarchy associated to a Frobenius manifold the density $f(u^1,\ldots,u^n)$ satisfies*

$$\frac{\partial^2 f}{\partial u^\alpha u^\beta} = c^\gamma_{\alpha\beta}\frac{\partial^2 f}{\partial u^\gamma \partial u^1}.$$

*If, moreover, $f(u^1,\ldots,u^n)$ is polynomial in $u^1$, then there exist $c^{\alpha,p}\in\mathbb{R}$ such that*

$$f(u) = \sum_{\alpha,p} c^{\alpha,p} h_{\alpha,p}(u)$$

*where only a finite number of constants $c^{\alpha,p}$ are nonzero.*

Next we notice that any dispersionless BHTS hierarchy is clearly invariant under the space-time rescaling $(x,t)\mapsto(cx,ct)$, where $c\in\mathbb{R}^*$. The infinitesimal generator of such a symmetry has the form

$$u^\gamma_c \;=\; \sum_{\alpha,p} u^\gamma_{t^{\alpha,p}} t^{\alpha,p} + u^\gamma_x x \;=\; \left(\sum_{\alpha,p} t^{\alpha,p}\eta^{\gamma\mu}\frac{\partial^2 h_{\alpha,p}}{\partial u^\mu \partial u^\nu} + x\delta^\gamma_\nu\right) u^\nu_x.$$

We can combine such a symmetry with any other symmetry $\frac{1}{2\pi}\int_0^{2\pi} f(u)\mathrm{d}x$ to obtain a new commuting flow

$$u^\alpha_s \;=\; u^\alpha_c + \eta^{\alpha\mu}\frac{\partial^2 f}{\partial u^\mu \partial u^\nu}u^\nu_x$$

and perform stationary reduction of the dynamics at $u^\alpha_s = 0$ to get a solution to our BHTS hierarchy in the implicit form

$$\sum_{\alpha,p} t^{\alpha,p}\eta^{\gamma\mu}\frac{\partial^2 h_{\alpha,p}}{\partial u^\mu \partial u^\nu} + x\delta^\gamma_\nu \;=\; \eta^{\alpha\mu}\frac{\partial^2 f}{\partial u^\mu \partial u^\nu}.$$

By the last theorem we conclude

**Theorem 1.6.** *Let $\frac{1}{2\pi}\int_0^{2\pi} f(u)\mathrm{d}x$ be any symmetry of the dispersionless BHTS hierarchy associated to a Frobenius manifold. Then a solution to the hierarchy can be found by* *solving for* $u(x,t)$ *the equation*

$$x\,\eta_{\gamma 1} + \sum_{\alpha,p} t^{\alpha,p}\frac{\partial h_{\alpha,p-1}(u)}{\partial u^\gamma} = \frac{\partial^2 f(u)}{\partial u^\gamma \partial u^1}.$$

*If* $h(u)$ *is polynomial in* $u^1$ *and* $h(u) = \sum_{\alpha,p} c^{\alpha,p} h_{\alpha,p}(u)$, *this is equivalent to*

$$x\,\eta_{\gamma 1} + \sum_{\alpha,p}(t^{\alpha,p} - c^{\alpha,p})\frac{\partial h_{\alpha,p-1}(u)}{\partial u^\gamma} = 0.$$

**Remark 1.9.** There is a special solution to a dispersionless BHTS hierarchy, identified by the initial datum $u^\alpha(x,t=0) = \delta^\alpha_1 x$ or, equivalently, by the choice

$$c^{\alpha,0} = \delta^\alpha_1, \qquad c^{\alpha,p} = 0,\ p > 0.$$

Such a solution is called a *topological solution* and it is easy to check that its $\tau$-function has the property

$$\epsilon^2 \log \tau_{\text{top}}(t)\Big|_{\substack{t^{\alpha,0}=u^\alpha \\ t^{\alpha,p}=0,\ p>0}} = F(u).$$

In the case of Frobenius manifolds coming from quantum cohomology the full topological $\tau$-function coincides with the full genus 0 Gromov-Witten potential with descendants

$$\mathbf{f}(t) = \epsilon^2 \log \tau_{\text{top}}(t).$$

This fact is equivalent to the validity of topological recursion relations for $\mathbf{f}(t)$, something true in general, and it is the genus 0 version of Witten's conjecture.

**Reconstruction of the dispersive tail.** We now briefly address the problem of classifying the possible dispersive deformations ($\epsilon \neq 0$) of a dispersionless BHTS hierarchy. This is important also in view of Dubrovin's reconstruction procedure for higher genera Gromov-Witten invariants starting from a semisimple rational quantum cohomology. In fact semisimplicity of the Frobenius structure plays a central role in the proof of the following main theorem. A Frobenius manifold is called *semisimple* if the Frobenius algebra at a point in each connected component of $U$ (and hence on an open, dense subset of $U$) is semisimple. A sufficient condition for semisimplicity is that the operator $(E\star)$ of multiplication by the Euler vector field has pairwise distinct eigenvalues.

**Theorem 1.7** ([8]). *All BHTS hierarchies coming from a semisimple Frobnius manifold can be reduced to dispersionless* ($\epsilon = 0$) *form by a* quasi-Miura *transformation*

$$u^\alpha \mapsto \tilde{u}^\alpha = u^\alpha + \sum_{k=1}^{\infty} \epsilon^k F^\alpha_k(u, u_x, u_{xx}, \ldots, u^{n_k})$$

*where* $F^\alpha_k(u, u_x, u_{xx}, \ldots, u^{n_k})$ *is a rational function in the derivatives of* $u$ *of degree* $k$ *with respect to the usual long wave limit grading.*

In other words, the quasi-Miura group acts transitively on the space of deformations of a given semisimple dispersionless BHTS hierarchy, hence providing the tool to control them.

**Example 1.10.** The quasi-Miura transformation reducing the KdV hierarchy to the dispersionless KdV hierarchy is given by (see [8])

$$\tilde{u} = u - \frac{\epsilon^2}{12}(\log u_x)_{xx} + \epsilon^4 \left[ \frac{u_{xxxx}}{288u_x^2} - \frac{7u_{xx}u_{xxx}}{480u_x^3} + \frac{u_{xx}^3}{90u_x^4} \right] + O(\epsilon^6).$$

**Remark 1.11.** In the case of semisimple quantum cohomologies, Dubrovin's scheme to reconstruct higher genera invariants starting from genus 0 consists in finding a specific quasi-Miura transformation for the BHTS hierarchy associated to the relevant semisimple Frobenius manifold and then formulating the following *generalized Witten's conjecture* ([8]) about the full Gromov-Witten potential with descendants at all genera

$$\mathbf{F}(t, \hbar) = \epsilon^2 \log \tau_{\text{top}}(t, \epsilon = \sqrt{\hbar}).$$

The privileged quasi-Miura transformation is selected using the so called *Virasoro symmetries* ([11]). We are not going to discuss them here and refer the reader to [8]. Roughly, they are an infinite sequence of extra non-local symmetries of the dispersionless BHTS hierarchy (they can be constructed using just the data of the Frobenius structure) and it is required that the quasi-Miura transformation preserve their expression as linear differential operators in the $t$ variables. This uniquely fixes the transformation, and hence the dispersive deformation of the BHTS hierarchy. Among the properties of such a trasformation is the fact that the above topological $\tau$-function depends only on even powers of $\epsilon$, so that $\mathbf{F}(t, \hbar)$ is a power series in the genus expansion variable $\hbar$. In the case of quantum cohomology of a point, Witten's conjecture ([18]) was proved by Kontsevich in [14].

## 2. Gromov-Witten Theory via Symplectic Field Theory

In this section we construct the dispersionless BHTS hierarchy associated to the quantum cohomology of a symplectic manifold without explicitly passing through Frobenius manifolds, but using instead Symplectic Field Theory. Comparing the formulas in this section with the ones in the previous section it will be evident how the two integrable systems coincide. Let $(M, \omega)$ be a closed symplectic manifold and, for simplicity, let us assume $H^i(M) = 0$ if $i$ is odd. We will consider Symplectic Field Theory of the framed Hamiltonian structure of fibration type given by the trivial bundle $V = S^1 \times M$. Remember that a Hamiltonian structure (see also [9]) is a pair $(V, \Omega)$, where $V$ is an oriented manifold of dimension $2n - 1$ and $\Omega$ a closed 2-form of maximal rank $2n - 2$. The line field $\text{Ker}\Omega$ is called the characteristic line field and we will call characteristic any vector field which generates $\text{Ker}\Omega$. A Hamiltonian structure is called stable if and only if there exists a 1-form $\lambda$ and a characteristic vector field $R$ (called Reeb vector field) such that

$$\lambda(R) = 1 \qquad \text{and} \qquad i_R \mathrm{d}\lambda = 0.$$

A framing of a stable Hamiltonian structure is a pair $(\lambda, J)$ with $\lambda$ as above and $J$ a complex structure on the bundle $\xi = \{\lambda = 0\}$ compatible with $\Omega$. In particular let $(M, \omega)$ be a closed symplectic manifold with a compatible almost complex structure $J_M$, $p : V \to M$ any $S^1$-bundle and $\lambda$ any $S^1$-connection form; then $(V, \Omega = p^*\omega, \lambda, J)$, with $J$ the lift of $J_M$ to the horizontal distribution, is a framed Hamiltonian structure.

In the case of $S^1 \times M$, a part of the invariants coming from holomorphic curve counting in $V \times \mathbb{R}$ coincide with those coming from holomorphic curve counting in $M$ and this correspondence will be enough to generate the same integrable system. Moreover the formalism of Symplectic Field Theory is more suitable to appreciate the topological meaning of the algebraic structure hidden in such systems.

## 2.1. SFT of Hamiltonian structures of fibration type and Gromov-Witten invariants

For now let $p : V \to M$ be any $S^1$-bundle over $M$. Recall that, if $J_M$ is an almost complex structure on $M$ compatible with $\omega$, $\lambda$ is any connection 1-form on V and the complex structure $J$ is any lift of $J_M$ to the horizontal distribution $\xi$, then $(V = S^1 \times M, \Omega = p^*\omega, \lambda, J)$ is a framed Hamiltonian structure.

Since each $S^1$-fibre of $V$ is a Reeb orbit for the Reeb vector field and we don't want to destroy the symmetry by perturbing this contact form to a generic one, we will need the non-generic Morse-Bott version of Symplectic Field Theory. Referring to [1] for the general construction, here we just stick to the case of fibrations, where the space $\mathcal{P}$ of periodic Reeb orbits can be presented as $\mathcal{P} = \coprod_{k=1}^{\infty} \mathcal{P}_k$, where each $\mathcal{P}_k$ is a copy of the base manifold $M$. Let then $\Delta_1, \dots, \Delta_b$ be a basis of $H^*(M)$ such that the system of forms $\tilde{\Delta}_j := p^*(\Delta_j)$, $j = 1, \dots, c \leq b$ generate $p^*(H^*(M)) \subset H^*(V)$, and the forms $\tilde{\Theta}_1, \dots, \tilde{\Theta}_d$ complete it to a basis of $H^*(V)$. Suppose $H_1(M) = 0$ and choose a basis $A_0, A_1, \dots, A_N$ of $H_2(M)$ in such a way that $\langle c_1(V), A_0 \rangle = l$ (if $l$ is the greatest divisor of the first Chern class $c_1(V)$ of our fibration), $\langle c_1(V), A_i \rangle = 0$, $i \neq 0$, and a basis of $H_2(V)$ is given by the lifts of $A_1, \dots, A_N$ if $l \neq 0$, $A_0, A_1, \dots, A_N$ if $l = 0$.

We consider a graded Poisson algebra $\mathfrak{P}$ formed by series in the formal variables $t^{1,j}, \dots, t^{c,j}$; $\tau^{1,j}, \dots, \tau^{d,j}$ $(j = 0, 1, \dots)$ associated to the string of forms $(\tilde{\Delta}_1, \dots, \tilde{\Delta}_c; \tilde{\Theta}_1, \dots, \tilde{\Theta}_d)$, the variables $p^1_k, \dots, p^b_k$ associated to the the classes $(\Delta_1, \dots, \Delta_b)$ and $\hbar$, with coefficients which are polynomials in the variables $q^1_k, \dots, q^b_k$. The Poisson structure is given in terms of the Poincaré pairing $\eta^{ij}$ in $H^*(M)$ as

$$\{p^i_k, q^j_l\} = k \delta_{k,l} \eta^{ij}.$$

Encoding the $p$ and $q$ variables into the the generating funcions

$$u^n(x) := \sum_{k=1}^{\infty} \left( q_k^n \, e^{-\mathrm{i}\, kx} + p_k^n \, e^{\mathrm{i}\, kx} \right) \qquad n = 1, \ldots, b,$$

such Poisson structure can be expressed as a formal distribution

$$\{u^i(x), u^j(y)\} = \eta^{ij} \delta'(x-y)$$

where $\eta^{ij}$ is the non-degenerate Poincaré pairing in $H^*(M)$. The grading of the variables is given by:

$$\begin{aligned} \deg(t^{ij}) &= 2(j-1) + \deg(\tilde{\Delta}_i) \\ \deg(\tau^{ij}) &= 2(j-1) + \deg(\tilde{\Theta}_i) \\ \deg(q_k^i) &= \deg(\Delta_i) - 2 + 2ck \\ \deg(p_k^i) &= \deg(\Delta_i) - 2 - 2ck \\ \deg(z_i) &= -2c_1(A_i) \end{aligned}$$

where $c = \frac{\langle c_1(TM), A_0 \rangle}{l}$ (see [10] for details on how to deal with fractional degrees).

The SFT-potential $\mathbf{h}$ is defined as a power series in the $t, \tau, p, q, z$ variables, where the coefficient of the monomial

$$t^{\alpha_1, i_1} \ldots t^{\alpha_r, i_r} \tau^{\beta_1, j_1} \ldots \tau^{\beta_s, j_s} p_{k_1^+}^{\alpha_1^+} \ldots p_{k_\mu^+}^{\alpha_\mu^+} q_{k_1^-}^{\alpha_1^-} \ldots q_{k_\nu^-}^{\alpha_\nu^-} z_1^{D_1} \ldots z_N^{D_N}$$

is given by the integral

$$\int_{\overline{\mathcal{M}}^A_{0,r+s,\mu,\nu}} \mathrm{ev}_1^* \tilde{\Delta}_{\alpha_1} \wedge \psi_1^{i_1} \wedge \ldots \wedge \mathrm{ev}_r^* \tilde{\Delta}_{\alpha_r} \wedge \psi_r^{i_r} \wedge \mathrm{ev}_{r+1}^* \tilde{\Theta}_{\beta_1} \wedge \psi_{r+1}^{j_1} \wedge \ldots \wedge \mathrm{ev}_{r+s}^* \tilde{\Theta}_{\beta_r} \wedge \psi_{r+s}^{j_r}$$
$$\wedge (\mathrm{ev}_1^+)^* \Delta_{\alpha_1^+}^{k_1^+} \wedge \ldots \wedge (\mathrm{ev}_\mu^+)^* \Delta_{\alpha_\mu^+}^{k_\mu^+} \wedge (\mathrm{ev}_1^-)^* \Delta_{\alpha_1^-}^{k_1^-} \wedge \ldots \wedge (\mathrm{ev}_\nu^-)^* \Delta_{\alpha_\nu^-}^{k_\nu^-}$$

where $\psi_i = c_1(L_i)$ is the $i$-th psi-class (like in Gromov-Witten theory, but see [12],[13] for a rigorous definition and further discussion in the SFT setting) and the integral is over the moduli space of holomorphic curves in $V \times \mathbb{R}$ with $r+s$ marked points and $\mu/\nu$ positive/negative punctures asymptotically cylindrical over Reeb orbits and realizing, together with the chosen capping surfaces (see [10] for details), the homology class $A = \sum D_i A_i$ in $H_2(V)$, modulo the $\mathbb{R}$ action coming from the $\mathbb{R}$ symmetry of the cylindrical target space $V \times \mathbb{R}$. The maps $\mathrm{ev}_i : \overline{\mathcal{M}}^A_{0,r+s,\mu,\nu} \to V$ are the evaluation map at the marked points, while the maps $(\mathrm{ev}_i^\pm) : \overline{\mathcal{M}}^A_{0,r+s,\mu,\nu} \to \mathcal{P}$ are the evaluation maps at the punctures, with $\Delta_\alpha^k$ the image of $\Delta_\alpha$ in $\mathcal{P}_k$.

Such potential is an element of $\mathfrak{P}$ of degree $-1$ and it satisfies the master equation

$$\{\mathbf{h}, \mathbf{h}\} = 0.$$

Let us consider the subalgebra $\tilde{\mathfrak{P}} \subset \mathfrak{P}$ obtained by setting to zero all the descendants of the $t$ variables, namely $t^{i,j} = 0$ for $j > 0$. One can expand $\mathbf{h}|_{t^{i,j}=0,j>0}$ in the number of $\tau$-variables

$$\mathbf{h}|_{t^{i,j}=0,j>0} = \mathbf{h}^0 + \sum_{i,j} \mathbf{h}^1_{i,j}\tau^{i,j} + \sum_{i_1,j_1,i_2,j_2} \mathbf{h}^2_{i_1,j_1;i_2,j_2}\tau^{i_1,j_1}\tau^{i_2,j_2} + \dots$$

and get sequences $\mathbf{h}^0$, $\mathbf{h}^1_{i,j}$, $\mathbf{h}^2_{i_1,j_1;i_2,j_2}$ of elements in the subalgebra $\tilde{\mathfrak{P}}^0 \subset \tilde{\mathfrak{P}} \subset \mathfrak{P}$ where all the $s$-variables and the descendants of the $t$-variables are set to zero.

Moreover, expanding in the same way the master equation $\{\mathbf{h}, \mathbf{h}\} = 0$ one gets that:

1) $d^0 = \{\mathbf{h}^0, \cdot\} : \tilde{\mathfrak{P}}^0 \to \tilde{\mathfrak{P}}^0$ makes $\tilde{\mathfrak{P}}^0$ into a differential algebra, since $d^0 \circ d^0 = 0$,
2) the Poisson bracket on $\tilde{\mathfrak{P}}^0$ descends to the homology $H_*(\tilde{\mathfrak{P}}^0, d^0)$, thanks to the Jacobi identity,
3) $\{\mathbf{h}^1_{i_1,j_1}, \mathbf{h}^1_{i_2,j_2}\} = 0$ as homology classes in $H_*(\tilde{\mathfrak{P}}^0, d^0)$.

However, because of the $S^1$ symmetry, we will see that $\mathbf{h}^0 = 0$, so that $H_*(\tilde{\mathfrak{P}}^0, d^0) = \tilde{\mathfrak{P}}^0$.

The main result of this section is the following

**Proposition 2.1** ([1],[10]). *Let $\mathbf{f}_M(\sum t^{i,j}\Delta_i, z)$ be the genus 0 Gromov-Witten potential of $M$ and $\mathbf{h}_V(\sum t^{i,j}\tilde{\Delta}_i + \sum \tau^{k,l}\tilde{\Theta}_k, q, p)$ the rational SFT potential of $V$ (as a framed Hamiltonian structure of fibration type). Let*

$$\mathbf{h}_{k,l}(t,q,p) = \frac{\partial \mathbf{h}_V}{\partial \tau^{k,l}}\left(\sum_1^c t^{i,j}\tilde{\Delta}_{ij} + \tau^{k,l}\tilde{\Theta}_k, q, p\right)\Bigg|_{\tau^{k,l}=0;\ t^{i,j}=0,j>0}$$

$$h_{k,l}(t;z) = \frac{\partial \mathbf{f}_M}{\partial \tau^{k,l}}\left(\sum_1^b t^{i,j}\Delta_i + \tau^{k,l}\pi_*\tilde{\Theta}_k, z\right)\Bigg|_{\tau^{k,l}=0;\ t^{i,j}=0,j>0}$$

*where $\pi_*$ denotes integration along the fibers of $V$. Then we have*

$$\mathbf{h}_{k,l}(t,q,p) = \frac{1}{2\pi}\int_0^{2\pi} h_{k,l}(t^{1,0} + u^1(x), \dots, t^{b,0} + u^b(x), u^{b+1}(x), \dots, u^c(x); \tilde{z})\mathrm{d}x$$

*where*

$$u^n(x) := \sum_{k=1}^{\infty} \left(q^n_k\, e^{-\mathrm{i}\, kx} + p^n_k\, e^{\mathrm{i}\, kx}\right) \qquad n = 1, \dots, b$$

*and $\tilde{z} = (\mathrm{e}^{-ilx}, z_1, \dots, z_N)$.*

*Proof.* The proof is based on a correspondence between the moduli spaces relevant for Gromov-Witten theory of $M$ and Symplectic Field Theory of $V$. Namely, let $p : L \to M$ be the complex line bundle associated to the $S^1$-bundle $V$ and let $\tilde{\mathcal{C}}^{\tilde{d};l_1,\dots,l_\alpha,m_1,\dots,m_\beta}_{r+1}(L)$ be the moduli space of pairs $(F = (f; x_1, \dots, x_{r+1}; x^+_1, \dots, x^+_\alpha; x^-_1, \dots, x^-_\beta), \sigma)$ where $F$ is a rational holomorphic curve in $M$ from the moduli space $\mathcal{M}^d_{g=0,r+1+\alpha+\beta}(M)$, $\sigma$ is a meromorphic section of the holomorphic line bundle $f^*(L)$ (with first Chern class $d_0 l$) such that the divisor of its poles and zeros equals $\sum_{i=1}^{\alpha} l_i x_i^+ - \sum_{i_1}^{\beta} m_i x_i^-$ and the degrees are given by $d = (d_0, d_1, \ldots, d_N)$ and $\tilde{d} = (d_1, \ldots, d_N)$. Then each holomorphic curve in $V \times \mathbb{R}$ with cylindrical ends which are asymptotic to periodic Reeb orbits can be identified with an element of $\tilde{\mathcal{C}}_{r+1}^{\tilde{d};l_1,\ldots,l_\alpha,m_1,\ldots,m_\beta}(L)$. Forgetting the section $\sigma$ provides a projection $\mathrm{pr} : \tilde{\mathcal{C}}_{r+1}^{\tilde{d};l_1,\ldots,l_\alpha,m_1,\ldots,m_\beta}(L)/\mathbb{R} \to \mathcal{C}_{r+1}^{\tilde{d};l_1,\ldots,l_\alpha,m_1,\ldots,m_\beta}(L)$, where $\mathcal{C}_{r+1}^{\tilde{d};l_1,\ldots,l_\alpha,m_1,\ldots,m_\beta}(L)$ is the moduli space of holomorphic curves from $\mathcal{M}_{g=0,r+1+\alpha+\beta}^{d}(M)$ that satisfy $[\sum_{i=1}^{\alpha} l_i x_i^+ - \sum_{i_1}^{\beta} m_i x_i^-] = f^*(L)$. Since a meromorphic section of $f^*(L)$ is determined by its divisor of poles and zeros up to a multiplicative complex constant, the fiber of pr is $S^1$. Actually it is easy to see that this $S^1$ bundle is isomorphic to the pull-back $ev_i^*(V)$ of the $S^1$-bundle $V$ through any of the $r+1$ evaluation maps $ev_i : \mathcal{C}_{r+1}^{\tilde{d};l_1,\ldots,l_\alpha,m_1,\ldots,m_\beta}(L) \to M$. Finally, the result follows by considering that $\mathcal{C}_{r+1}^{\tilde{d};l_1,\ldots,l_\alpha,m_1,\ldots,m_\beta}(L) = \mathcal{M}_{g=0,r+1+\alpha+\beta}^{d}(M)$ if $\sum l_i - \sum m_j = d_0 l$ and $\mathcal{C}_{r+1}^{\tilde{d};l_1,\ldots,l_\alpha,m_1,\ldots,m_\beta}(L) = \varnothing$ otherwise, since holomorphic line bundles on $\mathbb{P}^1$ are classified by their degree. The theorem, in fact, relates integration over $\tilde{\mathcal{C}}_{r+1}^{\tilde{d};l_1,\ldots,l_\alpha,m_1,\ldots,m_\beta}(L)/\mathbb{R}$ and over $\mathcal{C}_{r+1}^{\tilde{d};l_1,\ldots,l_\alpha,m_1,\ldots,m_\beta}(L)$. Integration over $x$ with the substitution $\tilde{z} = (\mathrm{e}^{-ilx}, z_1, \ldots, z_N)$ imposes the condition $\sum l_i - \sum m_j = d_0 l$. Finally notice that the single fiber class $\tilde{\Theta}_k$ serves to fix the remaining $S^1$-symmetry since the Poincaré dual of $\tilde{\Theta}_k$ is transverse to the fibers of $V$. □

Notice that, thanks to the $S^1$-symmetry of the Hamiltonian structure, one can deduce a priori that $\mathbf{h}^0 = 0$, so that $H_*(\tilde{\mathfrak{P}}, d^0) = \tilde{\mathfrak{P}}$.

**Remark 2.1.** The above result can be easily generalized (see [16]) to the case where $V$ is the total space of an $S^1$-orbibundle over a closed symplectic orbifold. In this case everything is to be rephrased in terms of orbifold cohomology and orbifold Gromov-Witten theory in the sense of [3],[4].

## 2.2. The trivial bundle case: Gromov-Witten theory

Let us now get back to the trivial bundle case $V = S^1 \times M$. Choose $\tilde{\Theta}_i = \tilde{\Delta}_i \wedge \mathrm{d}\phi$, $i = 1, \ldots, b$, where $\phi$ is the angle variable on $S^1$ and $b$ denotes the dimension of $H^*(M)$. Moreover let $A_1, \ldots, A_N$ be a basis in $H_2(M)$. In this case the Hamiltonians appearing in Theorem 2.1, in terms of the Gromov-Witten potential of $M$, are given by

$$\mathbf{h}_{k,l}(t,q,p) = \frac{1}{2\pi}\int_0^{2\pi} h_{k,l}(v^1(x), \ldots, v^b(x))\mathrm{d}x$$

with $v^i(x) = t^i + u^i(x)$ and

$$h_{k,l} = \left.\frac{\partial \mathbf{f}_M}{\partial t^{k,l}}\right|_{t^{i,j}=0, j>0}$$

Recall that, by definition of Gromov-Witten potential, $\eta_{ij} = \partial_1 \partial_i \partial_j \mathbf{f}_M$ and denote $t^i = t^{i,0}$, $\mathbf{h}_i = \mathbf{h}_{i,0}$ and $h_i(v) = h_{i,0}(v)$. Denote moreover $c_{ijk} = \partial_i \partial_j \partial_k \mathbf{f}_M$. We now show that $c_{ij}^k$ (where we use the metric $\eta$ to raise and lower the indices) are structure constants of a commutative associative algebra on each tangent space $T_tH^*(M)$. Notice preliminarily that, if $F(t,p,q) = \frac{1}{2\pi}\int_0^{2\pi} f(v(x))\mathrm{d}x$ and $G(t,p,q) = \frac{1}{2\pi}\int_0^{2\pi} g(v(x))\mathrm{d}x$, then

$$\{F,G\} = \frac{1}{2\pi}\int_0^{2\pi} \partial_i f \eta^{ij} \partial_x \partial_j g \ \mathrm{d}x$$

hence

$$\{F,G\} = 0 \ \Leftrightarrow \ \exists\ \Omega = \Omega(v) \mid \partial_i f \eta^{ij} \partial_x \partial_j g = \partial_x \Omega.$$

This is in turn equivalent to exactness of the 1-form $\partial_i f \eta^{ij} \partial_x \partial_j g \mathrm{d}v^k$ and hence amounts to

$$\partial_j \partial_i f \partial_k \partial^i g = \partial_k \partial_i f \partial_j \partial^i g. \tag{2}$$

Commutativity follows trivially by definition of $c_{ij}^k$. Associativity is equivalent to the Gromov-Witten potential $\mathbf{f}_M(t)$ of $M$ satisfying WDVV associativity equations

$$\partial_\alpha \partial_k \partial^i \mathbf{f}_M \partial_i \partial_\beta \partial_j \mathbf{f}_M = \partial_k \partial_\beta \partial^i \mathbf{f}_M \partial_\alpha \partial_i \partial_j \mathbf{f}_M. \tag{3}$$

This is easily verified using the equations $\{\mathbf{h}_\alpha, \mathbf{h}_\beta\} = 0$ and (2).

Exactly in the same way, but now considering the equations $\{\mathbf{h}_\alpha, \mathbf{h}_{\beta,j}\} = 0$ we get

$$c_{\alpha i}^k \partial_k \partial_s h_{\beta,j} = c_{\alpha s}^k \partial_i \partial_k h_{\beta,j}$$

and, for $i = 1$,

$$\partial_\alpha \partial_s h_{\beta,j} - c_{\alpha s}^k \partial_k \partial_1 h_{\beta,j} = 0. \tag{4}$$

These last equations, together with WDVV equations (3), are actually equivalent to commutativity of the whole system of Hamiltonians $\mathbf{h}_{\alpha,j}$ as the follwing simple Lemma shows.

**Lemma 2.2.**

$$\{\mathbf{h}_{\alpha,i}, \mathbf{h}_{\beta,j}\} = 0 \Leftrightarrow \begin{cases} \{\mathbf{h}_\alpha, \mathbf{h}_\beta\} = 0 \\ \partial_\alpha \partial_s h_{\beta,j} - c_{\alpha s}^k \partial_k \partial_1 h_{\beta,j} = 0. \end{cases}$$

*Proof.* We only need to prove the $\Leftarrow$ part. Using $\partial_\alpha \partial_s h_{\beta j} - c_{\alpha s}^k \partial_k \partial_1 h_{\beta j} = 0$ we get

$$\partial_\nu \partial_\gamma h_{\alpha i} \partial^\gamma \partial_\delta h_{\beta j} - \partial_\nu \partial_\gamma h_{\beta j} \partial^\gamma \partial_\delta h_{\alpha i} = c_{\nu\gamma}^\epsilon c_\delta^{\gamma\mu} (\partial_1 \partial_\epsilon h_{\alpha i} \partial_1 \partial_\mu h_{\beta j} - \partial_1 \partial_\mu h_{\alpha i} \partial_1 \partial_\epsilon h_{\beta j})$$

which vanishes thanks to associativity. This is equivalent to $\{\mathbf{h}_{\alpha,i}, \mathbf{h}_{\beta,j}\} = 0$. □

Finally we need to put into play the information about the grading of our generating functions. In particular we know that, with the above choice of degrees for the involved variables, the Gromov-Witten potential has degree $2(n-3)$ if $2n$ is the dimension of $M$. This can be expressed by saying that there exists an Euler vector field, linear in the variables $t$ and $z$, of the form:

$$\tilde{E} = \sum_{\alpha=1}^{b} (-2 + \deg \Delta_\alpha)\, t^\alpha \frac{\partial}{\partial t^\alpha} - \sum_{i=1}^{N} 2c_1(A_i) z_i \frac{\partial}{\partial z_i}$$

and, recalling the definition of the Hamiltonian densities $h_{\alpha,j}$, such that

$$\tilde{E}(h_{\alpha,j}) = \{[2(n-3)] - [2(j-1) + \deg \Delta_\alpha]\}\, h_{\alpha,j}. \tag{5}$$

In order to proceed further we will need some properties and recursion relations for Gromov-Witten invariants. We will state them here, without proving them, in terms of the Gromov Witten correlators with descendants

$$\langle \Delta_{i_1}\psi^{j_1}, \ldots, \Delta_{i_n}\psi^{j_n}\rangle_g^A = \int_{\bar{\mathcal{M}}_{g,n}^A(M)} \mathrm{ev}_1^*(\Delta_{i_1}) \wedge \psi_1^{j_1} \wedge \ldots \wedge \mathrm{ev}_n^*(\Delta_{i_n}) \wedge \psi_n^{j_n}$$

where, as usual, $g$ is the genus, $n$ the number of marked points and $A \in H_2(M)$ is the degree of the curves in the moduli space $\bar{\mathcal{M}}_{g,n}^A(M)$. We address the reader to [2] for a clear discussion and proof.

The first one is called *fundamental class axiom* and it takes the form

$$\langle \Delta_{i_1}\psi^{j_1}, \ldots, \Delta_{i_{n-1}}\psi^{j_{n-1}}, 1\rangle_g^A = \sum_{k=1}^{n-1} \langle \Delta_{i_1}\psi^{j_1}, \ldots, \Delta_{i_{k-1}}\psi^{j_{k-1}}, \Delta_{i_k}\psi^{j_k - 1}, \ldots, \Delta_{i_{n-1}}\psi^{j_{n-1}}\rangle_g^A$$

The direct consequence we need is easily translated into the following relation among Hamiltonians

$$\partial_1 h_{\alpha,j} = h_{\alpha,j-1}.$$

This equation can be plugged into equation (4) to get

$$\partial_\alpha \partial_s h_{\beta,j} - c_{\alpha s}^k \partial_k h_{\beta,j-1} = 0. \tag{6}$$

The second axiom we need is the *divisor axiom* and, if $D \in H^2(M)$ and either $D \neq 0$ or $n \geq 4$, it spells out as

$$\begin{aligned} \langle \Delta_{i_1}\psi^{j_1}, \ldots, \Delta_{i_{n-1}}\psi^{j_{n-1}}, D\rangle_g^A &= \left(\int_A D\right) \langle \Delta_{i_1}\psi^{j_1}, \ldots, \Delta_{n-1}\psi^{j_{n-1}}\rangle_g^A \\ &+ \sum_{k=1}^{n-1} \langle \Delta_{i_1}\psi^{j_1}, \ldots, \Delta_{i_{k-1}}\psi^{j_{k-1}}, (D \cup \Delta_{i_k})\psi^{j_k - 1}, \ldots, \Delta_{i_{n-1}}\psi^{j_{n-1}}\rangle_g^A \end{aligned} \tag{7}$$

First of all notice that this means our generating functions depend polynomially on all the $t$ variables (by degree considerations) but those associated with 2-forms (which have degree 0). If $D = \sum \bar{t}_i \Delta_i$ denotes the generic 2-form, then $h_{\alpha,j}$ depends polynomially on $z^A e^{(D,A)}$ up to a further polynomial dependence on the $\bar{t}_i$'s, corresponding to the sum in the right hand side of the divisor axiom, which involves the classical cup product and lower descendants. Notice that the $\bar{t}_i$'s also appear polynomially in the term accounting for degree 0 curves with three marked points, which has the form $\int_M t^{\wedge 3}$. These considerations can be promptly translated in terms of generating functions by defining an Euler vector field on $H^*(M)$ (i.e. one which doesn't involve the $z$ variables anymore) of the form

$$E = \sum_{\alpha=1}^{b} [(-2 + \deg \Delta_\alpha)\, t^\alpha + r^\alpha]\, \frac{\partial}{\partial t^\alpha}$$

where $r^\alpha = -2c_1(\Delta_\alpha)$. This Euler vector field assigns degree $-2c_1(\Delta_\alpha)$ to $\mathrm{e}^{t_\alpha}$ when $\Delta_\alpha \in H^2(M)$ and, in light of the above considerations, we can rewrite the homogeneity condition (5) as

$$E(h_{\alpha,j}) - r^i\, {c_i}^k{}_\alpha|_{t=0}\, h_{k,j-1} = \{[2(n-3)] - [2(j-1) + \deg \Delta_\alpha]\}\, h_{\alpha,j} \qquad (8)$$

where ${c_i}^k{}_\alpha|_{t=0}$ are the structure constants of the classical cup product in $H^*(M)$ and, of course, the term which involves them is the correction term needed to take into account the further polynomial dependence of $h_{\alpha,j}$ on $t_\alpha$ when $\Delta_\alpha \in H^2(M)$ (both in degree 0 and $n = 3$, using $h_{\alpha,-1} = \sum \eta_{\alpha\beta} t_\beta$, and in the other cases as discribed by the divisor axiom).

It can be shown (see [8],[6]) that equations (6) and (8) completely determine the hamiltonians $h_{\alpha,j}$ by recursion, starting from $h_{\alpha,-1} = \sum \eta_{\alpha\beta} t_\beta$. In fact, if we define the generating functions of the hamiltonian densities as

$$h_\alpha(t,\zeta) = \sum_{k=0}^{\infty} h_{\alpha,k-1}(t)\zeta^k = \sum_A \langle \frac{\Delta_\alpha}{1-\zeta\psi}, t, \ldots, t\rangle_0^A z^A$$

and form the so called $J$-function

$$J_\alpha(t,\zeta) = h_\alpha(t,\zeta)\, \zeta^{\mathcal{V}}\, \zeta^R$$

where the matrix $\mathcal{V}$ is given by $\mathcal{V}^\gamma_\epsilon = \frac{2-n}{2}\delta^\gamma_\epsilon + \frac{1}{2}\partial_\epsilon E^\gamma$ and $R$ is the matrix of (classical) multiplication by the first Chern class $R^\gamma_\epsilon = -\frac{1}{2} r^i\, {c_i}^\gamma{}_\epsilon|_{t=0}$, then the equations above are equivalent to $J_\alpha$ being flat coordinates for the already mentioned *deformed flat connection* on $H^*(M) \times \mathbb{C}^*$

$$\tilde{\nabla}_X Y = X^\alpha \partial_\alpha Y + \zeta X \star Y$$
$$\tilde{\nabla}_{\frac{\mathrm{d}}{\mathrm{d}\zeta}} Y = \frac{\mathrm{d}}{\mathrm{d}\zeta} Y - \frac{1}{2} E \star Y - \frac{1}{\zeta}\mathcal{V} Y$$

where $X(\zeta), Y(\zeta) \in TH^*(M)$ and $\zeta \in \mathbb{C}^*$. Notice that, beacause of the grading definition in Symplectic Field Theory [10], our Euler vector Field $E$ (and hence also the operator $R$) differs by a $-\frac{1}{2}$ multiplicative factor from Dubrovin's one [8], which we defined above and which is more commonly used in the literature.

## References

[1] F. Bourgeois, *A Morse-Bott approach to Contact Homology*, PhD dissertation, Stanford University, 2002.

[2] D. A. Cox, S. Katz, *Mirror Symmetry and Algebraic Geometry*, AMS Mathematical Surveys and Monographs, Vol. 68.

[3] W. Chen, Y. Ruan, *A New Cohomology Theory of Orbifold* arXiv preprint arXiv:math/0004129v3 [math.AG]
[4] W. Chen, Y. Ruan, *Orbifold Gromov-Witten Theory*, arXiv preprint arXiv:math/0103156v1 [math.AG]
[5] B. Dubrovin, *Geometry of 2D topologica field theories*, Integrable Systems and Quantum Groups, Springer Lecture Notes in Math., vol. 1620 (1996), pp. 120 - 348.
[6] B.Dubrovin, *Painlevé transcendents and topological field theory*, The Painlevé property: one century later, R.Conte (Ed.), Springer Verlag, 1999, p.287-412.
[7] B. Dubrovin, S.P. Novikov, *On Poisson brackets of hydrodynamic type*, Soviet Math. Dokl. 279:2 (1984), 294-297.
[8] B. Dubrovin, Y. Zhang, *Normal forms of hierarchies of integrable PDEs, Frobenius manifolds and Gromov-Witten invariants*, preprint arXiv:math.DG/0108160.
[9] Y. Eliashberg, *Symplectic Field Theory and its applications*, Proceedings of the International Congress of Mathematicians, Madrid, Spain, 2006.
[10] Y. Eliashberg, A. Givental and H. Hofer, *Introduction to Symplectic Field Theory*, GAFA 2000 Visions in Mathematics special volume, part II, 560-673.
[11] T. Eguchi, K. Hori, and C.-S. Xiong, *Quantum cohomology and Virasoro algebra*, Phys. Lett. B402 (1997), 71-80.
[12] O. Fabert, *Gravitational descendants in Symplectic Field Theory*, arXiv:0907.0789.
[13] O. Fabert, P. Rossi, *Topological recursion relations in SFT*, in preparation.
[14] M. Kontsevich, *Intersection Theory on the moduli space of curves and the matrix Airy function*, Comm. Math. Phys. 147 (1992), no.1, 1-23.
[15] F. Magri, *A simple construction of integrable systems*, J. Math. Phys. (1978), 11561162.
[16] P. Rossi, *Gromov-Witten theory of orbicurves, the space of tri-polynomials and Symplectic Field Theory of Seifert fibrations*, preprint arXiv:0808.2626.
[17] S.P.Tsarev, *Geometry of hamiltonian systems of hydrodynamic type. The generalised hodograph method*, Mathematics in the USSR-Izvestiya, 1991, v.377, No. 2, p. 397-419.
[18] E. Witten, *Two-dimensional gravity and intersection theory on moduli space*, Surveys in differential geometry (Cambridge MA, 1990), 243-310, Lehigh Univ. Bethlehem, PA, 1991.

Ecole Polytechnique, Paris and MSRI, Berkeley
*E-mail address*: issoroloap@gmail.com

Proceedings of $16^{th}$ Gökova
Geometry-Topology Conference
pp. 58 – 98

# Moduli spaces of surfaces and monodromy invariants

*Fabrizio Catanese, Michael Lönne, Bronislaw Wajnryb*

Abstract. Bidouble covers $\pi : S \to Q := \mathbb{P}^1 \times \mathbb{P}^1$ of the quadric and their natural deformations are parametrized by connected families $\mathfrak{N}_{a,b,c,d}$ depending on four positive integers $a, b, c, d$. We shall call these surfaces abcd-surfaces. In the special case where $b = d$ we call them abc-surfaces.

Such a Galois covering $\pi$ admits a small perturbation yielding a general 4-tuple covering of $Q$ with branch curve $\Delta$, and two natural Lefschetz fibrations obtained from a small perturbation of the composition $p_i \circ \pi$, $i = 1, 2$, $p_i$ being the i-th projection of $Q$ onto $\mathbb{P}^1$.

The first and third author showed that the respective mapping class group factorizations corresponding to the first Lefschetz fibration are equivalent for two abc-surfaces with the same values of $a + c, b$, a result which implies the diffeomorphism of two such surfaces.

We report on a more general result of the three authors implying that the first braid monodromy factorization corresponding to $\Delta$ determines the three integers $a, b, c$ in the case of abc-surfaces. We provide in this article a new proof of the non equivalence of two such factorizations for different values of $a, b, c$.

We finally show that, under certain conditions, although the first Lefschetz fibrations are equivalent for two abc-surfaces with the same values of $a + c, b$, the second Lefschetz fibrations need not be equivalent.

These results rally around the question whether abc-surfaces with fixed values of $a + c, b$, although diffeomorphic but not deformation equivalent, might be not canonically symplectomorphic.

Contents

# 1. Introduction: Moduli spaces of surfaces of general type: deformation, differentiable and symplectic types.

Let $X$ be a smooth complex projective variety, i.e., $X$ is a smooth compact complex submanifold of $\mathbb{P}^n := \mathbb{P}^n_{\mathbb{C}}$ of complex dimension $d$, and $X$ is connected.

Observe that $X$ is a compact oriented real manifold of real dimension $2d$, so, for instance, a complex surface gives rise to a real 4-manifold.

For $d = 1$, $X$ is a complex algebraic curve (a real surface, called also Riemann surface) and its basic invariant is the genus $g = g(X)$, defined as the dimension of the vector space $H^0(\Omega^1_X)$ of holomorphic 1-forms.

The situation for curves is 'easy', since the genus determines the topological and the differentiable manifold underlying $X$: its intuitive meaning is the 'number of handles' that one must add to a sphere in order to obtain $X$ as a topological space.

The rough classification of curves is the following :

- $g = 0$ : $X \cong \mathbb{P}^1_{\mathbb{C}}$, topologically $X$ is a sphere $S^2$ of real dimension 2.
- $g = 1$ : $X \cong \mathbb{C}/\Gamma$ , with $\Gamma$ a discrete subgroup $\cong \mathbb{Z}^2$: $X$ is called an elliptic curve, and topologically we have a real 2-torus $S^1 \times S^1$.
- $g \geq 2$ : then we have a 'curve of general type', and topologically we have a connected sum of $g$ 2-tori.

Every curve is a deformation of a special hyperelliptic curve, the non singular projective model of the affine curve

$$\{(x, z) \in \mathbb{C}^2 | z^2 = (x-1)(x-2)\dots(x-(2g+2))\}.$$

The main theorem for curves says more precisely that we have a **Moduli space** $\mathfrak{M}_g$ which parametrizes the isomorphism classes $[C]$ of compact complex curves $C$ of genus $g$.

$\mathfrak{M}_g$ is a Zariski open set of a complex projective variety, is connected, and has complex dimension $(3g-3) + a(g)$, where $a(g)$ is the complex dimension of the group of complex automorphisms $(a(0) = 3, a(1) = 1, a(g) = 0, \ for \ g \geq 2)$.

When we pass to complex dimension $d = 2$, some features do generalize, others do not.

The first generalizations of the genus were given by Clebsch, Noether, Enriques and Castelnuovo through the dimensions of certain vector spaces of holomorphic tensor differential forms

$$q(X) := dim_{\mathbb{C}} H^0(\Omega^1_X), p_g(X) := dim_{\mathbb{C}} H^0(\Omega^d_X), P_m(X) := dim_{\mathbb{C}} H^0(\Omega^d_X{}^{\otimes m}).$$

$q(X)$ is called the irregularity of $X$, $p_g(X) = P_1(X)$ is called the geometric genus, while $P_m(X)$ is called the m-th plurigenus.

These invariants suffice to give the Castelnuovo-Enriques classification of algebraic surfaces.

The rough classification of projective surfaces $S$ is the following :

- $P_{12} = 0$ : $S$ is birational to a product $C \times \mathbb{P}^1_{\mathbb{C}}$, where $C$ is a complex curve of genus $q(S)$.

- $P_{12} = 1$: the surface is birational to a surface $S$ which is an analogue of an elliptic curve, in the sense that there is a holomorphic section of $H^0(\Omega_S^{2\otimes 12})$ which is nowhere vanishing.
- $P_{12} \geq 2$ and $S$ is properly elliptic, i.e., the rational map $\phi_{12}$ associated to the sections of $H^0(\Omega_S^{2\otimes 12})$ maps to a curve and the general fibres are elliptic curves.
- $P_{12} \geq 2$ and $S$ is of general type, i.e., $\phi_{12}$ maps birationally to a surface $X$, called the canonical model of $S$.

Surfaces of special type are well understood, but surfaces of general type still offer us a lot of intriguing and fascinating open problems. In each birational class there is for them a unique smooth projective surface (up to isomorphism) $S$, called minimal model, with the property that every birational holomorphic map $S$ onto another smooth surface $S'$ is a biholomorphism. This minimal model offers therefore also a unique topological and differentiable model. As does the genus of an algebraic curve, the above numerical invariants are determined by the topological structure of $S$:

- $2q(S)$ is the first Betti number of $S$, $2q(S) = b_1(S)$, and
- if one defines $\chi(S) := 1 - q(S) + p_g(S)$, we have Noether's formula $12\chi(S) = K_S^2 + e(S)$, where $e(S) = c_2(S)$ is the topological Euler Poincaré characteristic, while $K_S^2 = c_1^2(S)$; moreover
- the signature $\sigma(S) := b^+(S) - b^-(S)$ of the intersection form $q_S : H^2(S, \mathbb{Z}) \to \mathbb{Z}$ equals, by virtue of Hodge's index theorem,

$$\sigma(S) = 1/3(K_S^2 - 2e(S))$$

  and finally we have Kodaira's Riemann-Roch formula (holding for surfaces of general type) which says that
- the plurigenus $P_m(S)$, for $m \geq 2$, equals $\frac{m(m-1)}{2} K_S^2 + \chi(S)$.

Things seem 'on the surface' to generalize nicely: because also for algebraic surfaces of general type there exist similar moduli spaces $\mathfrak{M}_{x,y}$ (by the results of [Bom73], [Gie77]). Here, $\mathfrak{M}_{x,y}$ is quasi projective and parametrizes isomorphism classes of minimal (smooth projective) surfaces of general type $S$ such that $\chi(S) = x, K_S^2 = y$.

The fact that $\mathfrak{M}_{x,y}$ is quasi-projective implies that $\mathfrak{M}_{x,y}$ has a finite number of connected components, which parametrize deformation classes of surfaces of general type; and, by a classical theorem of Ehresmann ([Ehr43]), deformation equivalent varieties are diffeomorphic.

Hence, fixed the two numerical invariants $\chi(S) = x, K_S^2 = y$, which are determined by the topology of $S$, we have a finite number of differentiable types, and a fortiori a finite number of topological types.

These two numbers $\chi(S), K_S^2$ are determined by the topology. But they do not determine the topology, since the fundamental group is not encoded in such invariants (and is not even invariant under the action of the absolute Galois group, cf. [Ser64], and [BCG07] for new examples).

For this reason one likes to restrict to the case of simply connected surfaces. In this case, these two numbers almost determine the topology. This follows by Michael Freedman's big Theorem of 1982 ([Free82]), showing that there are indeed at most two topological structures if the surface $S$ is assumed to be simply connected.

Topologically, our 4-manifold is then obtained from very simple building blocks, one of them being the K3 surface, where:

**Definition 1.1.** *A* **K3** *surface is a smooth surface of degree 4 in* $\mathbb{P}^3_{\mathbb{C}}$.

Now, a complex manifold carries a natural orientation corresponding to the complex structure, and, in general, given an oriented differentiable manifold $M$, $M^{opp}$ denotes the same manifold, but endowed with the opposite orientation. This said, one can explain the corollary of Freedman's theorem for the topological manifolds underlying simply connected (compact) complex surfaces as follows:

- if $S$ is EVEN, i.e., the intersection form on $H^2(S, \mathbf{Z})$ is even: then $S$ is homeomorphic to a connected sum of copies of $\mathbb{P}^1_{\mathbb{C}} \times \mathbb{P}^1_{\mathbb{C}}$ and copies of a $K3$ surface if the signature $\sigma(S)$ is negative, and of copies of $\mathbb{P}^1_{\mathbb{C}} \times \mathbb{P}^1_{\mathbb{C}}$ and of copies of a $K3$ surface with reversed orientation if the signature is positive.
- if $S$ is ODD : then $S$ is a connected sum of copies of $\mathbb{P}^2_{\mathbb{C}}$ and of ${\mathbb{P}^2_{\mathbb{C}}}^{opp}$.

Kodaira and Spencer defined quite generally ([K-S58]) for compact complex manifolds $X, X'$ the equivalence relation called **deformation equivalence**: this, for surfaces of general type, means that the corresponding isomorphism classes yield points in the same connected component of the moduli space $\mathfrak{M}$.

The cited theorem of Ehresmann guarantees that DEF $\Rightarrow$ DIFF: indeed a bit more holds, namely, deformation equivalence implies the existence of a diffeomorphism carrying the canonical class $K_X$ to the canonical class $K_{X'}$.

In the 80' s the work of Simon Donaldson ([Don83], [Don86],[Don90],[Don92], see also [D-K90]) showed that homeomorphism and diffeomorphism differ drastically for projective surfaces.

The introduction of Seiberg-Witten invariants (see [Wit94],[Don96],[Mor96]) showed then more easily that a diffeomorphism $\phi : S \to S'$ between minimal surfaces of general type satisfies $\phi^*(K_{S'}) = \pm K_S$.

Contrary to the conjecture made by the first author in [Kat83], Friedman and Morgan made the

FRIEDMAN-MORGAN'S SPECULATION (1987) (see [F-M88]):
differentiable and deformation equivalence coincide for surfaces.
We abbreviate this speculation with the acronym DEF = DIFF.

The question was answered negatively in every possible way ([Man01], [KK02], [Cat03], [CW04], [BCG07], see also [Cat08] for a rather comprehensive survey):

**Theorem 1.2.** *(Manetti 1998, Kharlamov-Kulikov 2001, Catanese 2001, Catanese-Wajnryb 2004, Bauer-Catanese-Grunewald 2005 )*

*The Friedman- Morgan speculation does not hold true.*

The counterexamples obtained by Catanese and Wajnryb are the only ones which are simply connected, and one also has (as also in Manetti's examples) non deformation equivalent surfaces $S, S'$ such that there exists an orientation preserving diffeomorphism $\phi : S \to S'$ with $\phi^*(K_{S'}) = K_S$.

The simply connected examples used, called **abc-surfaces**, are a special case of a class of surfaces which the first author introduced in 1982 ([Cat84]), namely, bidouble covers of the quadric and their natural deformations.

We shall briefly describe these surfaces in the next section.

But, before, let us explain why we believe that these surfaces deserve further attention. Recall the

**Definition 1.3.** *A pair $(X, \omega)$ of a real manifold $X$, and of a real differential 2- form $\omega$ is called a* **Symplectic pair** *if*

*i) $\omega$ is a symplectic form, i.e., $\omega$ is closed ($d\omega = 0$) and $\omega$ is nondegenerate at each point (thus $X$ has even real dimension).*

*A symplectic pair $(X, \omega)$ is said to be* **integral** *iff the De Rham cohomology class of $\omega$ comes from $H^2(X, \mathbb{Z})$, or, equivalently, there is a complex line bundle $L$ on $X$ such that $\omega$ is a first Chern form of $L$.*

*ii) An almost complex structure $J$ on $X$ (a differentiable endomorphism of the real tangent bundle of $X$ satisfying $J^2 = -1$) is said to be compatible with $\omega$ if $\omega(Jv, Jw) = \omega(v, w)$, and the quadratic form $\omega(v, Jv)$ is strictly positive definite.*

For long time (before the examples of Kodaira and Thurston) the basic examples of symplectic manifolds were given by symplectic submanifolds of Kähler manifolds, in particular of the flat space $\mathbb{C}^n$ and of projective space $\mathbb{P}^N_{\mathbb{C}}$, endowed with the Fubini-Study form

$$\omega_{FS} := \frac{i}{2\pi} \partial\bar{\partial} log|z|^2.$$

It was observed recently by the first author ([Cat02], [Cat06]):

**Theorem 1.4.** *A minimal surface of general type $S$ has a symplectic structure $(S, \omega)$, unique up to symplectomorphism, and invariant for smooth deformation, with class$(\omega) = K_S = -c_1(S)$.*

*This symplectic structure is called the* **canonical symplectic structure**.

The above result is, in the case where $K_S$ is ample, a rather direct consequence of the well known Moser's lemma ([Mos65]), since then it suffices to pull-back $1/m$ of the Fubini-Study metric by an embedding of $S$ by the sections of $H^0(S, \Omega_S^{2\,\otimes m})$.

It is the case where $K_S$ is not ample which needs a non trivial proof.

An important consequence of the techniques of symplectic approximation of singularites employed in these papers is the following theorem ([Cat02], [Cat09], see also [Cat08] for a survey including the basics concerning the construction of the Manetti surfaces)

**Theorem 1.5.** *Manetti's surfaces yield examples of surfaces of general type which are not deformation equivalent but are canonically symplectomorphic.*

The following questions are still wide open.

**Questions:** 1) Are there (minimal) surfaces of general type which are orientedly diffeomorphic through a diffeomorphism carrying the canonical class to the canonical class, but, endowed with their canonical symplectic structure, are not canonically symplectomorphic?

2) Are there such simply connected examples?

3) Are there diffeomorphic abc-surfaces which are canonically symplectomorphic, thus yielding a counterexample to Can. Sympl = Def also in the simply connected case? Are there diffeomorphic abc-surfaces which are not canonically symplectomorphic, thus providing an answer to Question 1)?

The difficult problem of understanding the canonical symplectic structures of abc-surfaces ([CLW]) is relying on the work of Auroux and Katzarkov ([A-K00]), who described some invariants of integral symplectic pairs of real dimension 4. We postpone to a later section the description of their method.

## 2. abcd-Surfaces and their moduli spaces

All the curves are deformations of hyperelliptic curves $C$ , and the nature of the double cover $\pi : C \to \mathbb{P}^1$ allows an easy understanding of the topology of $C$.

Double covers of the projective plane $\mathbb{P}^2$ or of $\mathbb{P}^1 \times \mathbb{P}^1$ are however too simple surfaces, with very special behaviour; while it turned out that bidouble covers of $\mathbb{P}^1 \times \mathbb{P}^1$, introduced in [Cat84], exhibited a quite non special behaviour concerning properties of the moduli spaces of surfaces of general type.

Bidouble covers of the quadric are smooth projective complex surfaces $S$ endowed with a (finite) Galois covering $\pi : S \to Q := \mathbb{P}^1 \times \mathbb{P}^1$ with Galois group $(\mathbb{Z}/2\mathbb{Z})^2$.

More concretely, they are defined by a single pair of equations

$$z^2 = f_{(2a,2b)}(x_0, x_1; y_0, y_1)$$
$$w^2 = g_{(2c,2d)}(x_0, x_1; y_0, y_1)$$

where $a, b, c, d \in \mathbb{N}^{\geq 3}$ and the notation $f_{(2a,2b)}$ denotes that $f$ is a bihomogeneous polynomial, homogeneous of degree $2a$ in the variables $x$, respectively of degree $2b$ in the variables $y$.

These surfaces are simply connected and minimal of general type, and they were introduced in [Cat84] in order to show that the moduli spaces $\mathfrak{M}_{\chi,K^2}$ of smooth minimal surfaces of general type $S$ with $K_S^2 = K^2, \chi(S) := \chi(\mathcal{O}_S) = \chi$ need not be equidimensional or irreducible (and indeed the same holds for the open and closed subsets $\mathfrak{M}^{00}_{\chi,K^2}$ corresponding to simply connected surfaces).

In fact, for the dimension of the moduli space $\mathfrak{M}_{\chi,K^2}$ one can only give lower and upper bounds, as shown by the followig theorem (cf. [Cat84])

**Theorem 2.1.** *The dimension of an irreducible component $\mathfrak{M}'_{\chi,K^2}$ of $\mathfrak{M}_{\chi,K^2}$ is subject to the following inequalities:*

$$10\chi - 2K^2 \leq dim(\mathfrak{M}'_{\chi,K^2}) \leq 10\chi + 3K^2 + 108.$$

And the deformations of bidouble covers of the quadric were used to show that the moduli spaces $\mathfrak{M}^{00}_{\chi,K^2}$ could have as many irreducible components of different dimensions as possible.

These irreducible components were determined as follows: given our four integers $a, b, c, d \in \mathbb{N}^{\geq 3}$, consider the so called natural deformations of these bidouble covers of *simple type* $(2a, 2b), (2c, 2d)$, defined by equations [1]

$$(***) \quad z_{a,b}^2 = f_{2a,2b}(x, y) + w_{c,d}\phi_{2a-c,2b-d}(x, y)$$
$$w_{c,d}^2 = g_{2c,2d}(x, y) + z_{a,b}\psi_{2c-a,2d-b}(x, y)$$

where f, g, $\phi$, $\psi$, are bihomogeneous polynomials , belonging to respective vector spaces of sections of line bundles (recall that we defined $Q := \mathbb{P}^1 \times \mathbb{P}^1$):
$f \in H^0(Q, \mathcal{O}_Q(2a, 2b))$,
$\phi \in H^0(Q, \mathcal{O}_Q(2a - c, 2b - d))$ and
$g \in H^0(Q, \mathcal{O}_Q(2c, 2d))$,
$\psi \in H^0(Q, \mathcal{O}_Q(2c - a, 2d - b))$.

In this way one defines an open[2] subset $\mathfrak{N}'_{a,b,c,d}$ of the moduli space, whose closure $\overline{\mathfrak{N}'}_{a,b,c,d}$ is an irreducible component of $\mathfrak{M}_{\chi,K^2}$, where $\chi = 1 + (a - 1)(b - 1) + (c - 1)(d - 1) + (a + c - 1)(b + d - 1)$, and $K^2 = 8(a + c - 2)(b + d - 2)$.

We record here for further use: $\chi - \frac{1}{8}K^2 = ab + cd$.

These calculations can be derived from the formula $\mathcal{O}_S(K_S) = \pi^*\mathcal{O}_Q(a+c-2, b+d-2)$, which was later used in [Cat86] to show that the index of divisibility $r(S)$ of $K_S$ in $H^2(S, \mathbb{Z})$ equals the $G.C.D.(a + c - 2, b + d - 2)$. This result showed that many of these irreducible components $\overline{\mathfrak{N}'}_{a,b,c,d}$ would belong to distinct connected components of the moduli space (since the divisibility index $r(S)$ is constant on a connected component).

The **abc-surfaces** are obtained as the special case where $b = d$, and the upshot of [CW04] is that, once the values of the integers $b$ and $a + c$ are fixed, one obtains diffeomorphic surfaces.

It is relevant to observe that the family $\mathfrak{N}'_{a,b,c,d}$ is contained as a dense open subset in a larger irreducible family $\mathfrak{N}_{a,b,c,d}$, where coverings of degree 4 of other rational surfaces $\mathbb{F}_{2h} := Proj(\mathcal{O}_{\mathbb{P}^1} \oplus \mathcal{O}_{\mathbb{P}^1}(2h))$ have also to be considered.

In [Cat84] it was shown that two irreducible components of the moduli space, such as $\overline{\mathfrak{N}'}_{a,b,c,d}$, even when they have the same dimension, are distinct as soon as the 4-tuples are not trivially equivalent by an obvious $(\mathbb{Z}/2)^2$-symmetry (exchanging the two factors of $\mathbb{P}^1 \times \mathbb{P}^1$, or the two branch divisors). In fact an irreducible component defines a surface over the generic point, and if two families coincide, then they should define isomorphic surfaces over the generic point. In particular, these two surfaces should be isomorphic as surfaces over the algebraic closure of the function field of the generic point.

[1] in the following formula, a polynomial of negative degree is identically zero.

[2] deformation theory shows that this set is open

In this particular case, the corresponding surfaces were shown to be non isomorphic since this isomorphism would yield an isomorphism of the canonical images, and the canonical images were shown to have non equivalent singularities.

In general, it is much more difficult to show that two irreducible components yield distinct connected components of the moduli space.

In practice, in order to show that a certain open set $\mathfrak{N}$ of the moduli space is a connected component, one has to prove that the set $\mathfrak{N}$ is also closed.

In order to do so one has to look at all families of surfaces parametrized by a smooth curve $T$, and such that all the surfaces $S_t$, $t \in T \setminus \{t_0\}$ belong to $\mathfrak{N}$: if one can show that also the limit surface $S_{t_0}$ stays in $\mathfrak{N}$ then $\mathfrak{N}$ is then closed, hence a connected component.

Of course, if two surfaces $S, S'$ are not diffeomorphic (e.g., if their divisibiity indices of the canonical class satisfy $r(S) \neq r(S')$), then they lie in different connected components.

Boris Moishezon proposed instead to use, as an invariant of a connected component $\mathfrak{N}$ of the moduli space, the geometry of pluricanonical projections, which we shall describe in the final section.

## 3. Diffeomorphisms of abc-surfaces and Lefschetz pencils

The main new result of [CW04] was the following

**Theorem 3.1. (Catanese-Wajnryb)** *Let $S$ be an $(a, b, c)$ - surface (i.e., an $(a, b, c, b)$ surface) and $S'$ be an $(a+1, b, c-1)$-surface. Moreover, assume that $a, b, c-1 \geq 2$. Then $S$ and $S'$ are diffeomorphic.*

This result was then put together with the following more technical result, in order to exhibit simply connected surfaces which are diffeomorphic but not deformation equivalent.

**Theorem 3.2. (Catanese-Wajnryb)** *Let $S$, $S'$ be simple bidouble covers of $\mathbb{P}^1 \times \mathbb{P}^1$ of respective types ((2a, 2b),(2c,2b), and (2a + 2k, 2b),(2c - 2k,2b), and assume*

- *(I) $a, b, c, k$ are strictly positive even integers with $a, b, c-k \geq 4$*
- *(II) $a \geq 2c+1$,*
- *(III) $b \geq c+2$ and either*
- *(IV1) $b \geq 2a+2k-1$ or*
  *(IV2) $a \geq b+2$*

*Then $S$ and $S'$ are not deformation equivalent.*

The second theorem uses techniques which have been developed in a series of papers by the first author and by Marco Manetti during a long period of time, and it belongs more to algebraic geometry than to geometry and topology.

The essential thrust is to show, under the above technical conditions, that $\mathfrak{N}_{a,b,c,b}$ is indeed a connected component of the moduli space.

The substantial inequality is the inequality (II), which guarantees that for all natural deformations one has $\psi \equiv 0$ (since a polynomial of negative degree is identically zero), i.e., the natural deformations have the following form

$$(***) \quad z_{a,b}^2 = f_{2a,2b}(x,y) + w_{c,b}\phi_{2a-c,b}(x,y)$$

$$w_{c,b}^2 = g_{2c,2b}(x,y).$$

The above equations show that the covering $\pi : S \to Q$ is an iterated double covering, i.e., we first take an intermediate double covering $Y$ determined by taking the square root $w$ of $g$, and then we obtain (through the quadratic equation for $z$) $S$ as a double covering $S \to Y$.

The main idea is that, while taking limits, we can recover a smooth rational surface $\mathbb{F}_{2h}$ as the limit of $Q = Y/(\mathbb{Z}/2) = (S/(\mathbb{Z}/2))/(\mathbb{Z}/2)$.

We mention here however that the following question is rather open in general:

**Question:** For which values of $a, b, c, d$ is $\mathfrak{N}_{a,b,c,d}$ a connected component of the moduli space?

A more detailed exposition for both theorems can be found in the Lecture Notes of the C.I.M.E. courses 'Algebraic surfaces and symplectic 4-manifolds' (see especially [Cat08]).

The above result in differential topology Theorem 3.1 is based instead on a refinement of Lefschetz theory obtained by Kas ([Kas80]).

This refinement allows us to encode the differential topology of a 4-manifold $X$ Lefschetz fibred over $\mathbb{P}^1_{\mathbb{C}}$ (i.e., $f : X \to \mathbb{P}^1_{\mathbb{C}}$ has the property that all the fibres are smooth and connected, except for a finite number which have exactly one nodal singularity) into an equivalence class of a factorization of the identity in the Mapping class group $\mathcal{M}ap_g$ of a compact curve $C_0$ of genus $g$.

Recall that the mapping class group, introduced by Max Dehn([Dehn38]) in the 30's, is defined for each orientable manifold $M$ as

$$\mathcal{M}ap(M) := Diff^+(M)/Diff^0(M),$$

where $Diff^+(M)$ is the group of orientation preserving diffeomorphisms, while $Diff^0(M)$ is the connected component of the identity, the so called subgroup of the diffeomorphisms which are **isotopic to the identity**.

In the case of a complex curve of genus $g$, Dehn found that the group was generated by the so called Dehn twists around simple closed loops $\Gamma$, which are diffeomorphisms equal to the identity outside of a neighbourhood of $\Gamma$, and rotate $\Gamma$ by the antipodal map.

In our situation, let $F : X \to \mathbb{P}^1_{\mathbb{C}}$ be a Lefschetz fibration, and let $b_1, \dots b_m$ be the critical values of $f$. For each $b_i$ the fibre $C_i := f^{-1}(b_i)$ has a nodal singularity $P_i$, and the local monodromy around $b_i$ is given, by a theorem of Picard and Lefschetz, by the Dehn twist around the vanishing cycle in a nearby smooth fibre $C_{b'_i}$.

If one fixes a base point $b_0$ in the base $\mathbb{P}^1_{\mathbb{C}}$, and simple paths $\gamma'_i$ not crossing each other from $b_0$ to $b'_i$, one obtains diffeomorphisms, well defined up to isotopy, of the standard fibre $C_0 := C_{b_0}$ with the smooth fibres $C_{b'_i}$, and then a sequence of Dehn twists $\tau_1, \dots \tau_m$.

Assuming that the paths $\gamma'_i$ are ordered in counterclockwise order, by adding a small circle around the point $b_i$ run counterclockwise starting from $b'_i$, one obtains a geometric basis $\gamma_i$ for the fundamental group

$$\pi(\mathbb{P}^1_{\mathbb{C}} \setminus \{b_1, \dots b_m\}, b_0) = \langle \gamma_1, \dots \gamma_m | \gamma_1 \cdot \dots \cdot \gamma_m = 1 \rangle.$$

By virtue of the relation $\gamma_1 \cdot \dots \cdot \gamma_m = 1$ we obtain a sequence of Dehn twists $\tau_1, \dots \tau_m$ on $C_0 = C_{b_0}$ whose product $\tau_1 \cdot \dots \cdot \tau_m$ equals the identity. We obtain a factorization of the identity in $\mathcal{M}ap_g$ once we fix a diffeomorphism of the fibre $C_0 = C_{b_0}$ with a fixed compact complex curve of genus $g$.

This means that this factorization is only defined up to simultaneous conjugation in the group $\mathcal{M}ap_g$, i.e., one can replace, for $\phi \in \mathcal{M}ap_g$, each $\tau_i$ by

$$\tau_i^{\phi} := \phi^{-1} \tau_i \phi.$$

Moreover, there is another ambiguity for the factorization, due namely to the choice of the generators $\gamma_1, \dots \gamma_m$ for the fundamental group: this ambiguity amounts to an action of the braid group in $m$ strings on the factorization, leading to the so called Hurwitz equivalence for factorizations, and which will be described in the next section.

Having said this, one can state the theorem of Kas:

**Theorem 3.3.** *Two Lefschetz fibrations $F_1, F_2$ of the same genus $g \geq 2$ are diffeomorphic (there is a diffeomorphism $\varphi$ with $F_1 = F_2 \circ \varphi$) iff the corresponding factorizations in the mapping class group $\mathcal{M}ap_g$ are equivalent via Hurwitz equivalence and simultaneous conjugation.*

*Moreover, if, for $j = 1, 2$, $S_j$ is a complex surface and $F_j : S_j \to \mathbb{P}^1$ is homotopic to a holomorphic map, then the diffeomorphism $\varphi$ sends the canonical class $-c_1(S_1)$ in $H^2(S_1, \mathbf{Z})$ to the canonical class $-c_1(S_2)$.*

*Proof.* We have only added a second part to the statement, based on the Seiberg Witten theorem mentioned above, and on the fact that $\varphi$ sends fibres to fibres in an orientation preserving fashion. Hence $\varphi$ is orientation preserving, sends the class of the fibre to the class of a fibre, $c_1(S_1)$ to $\pm c_1(S_2)$: we conclude that the sign is $+1$, since the intersection number of a fibre with $-c_1(S_j)$ equals $2g - 2 > 0$.

*Q.E.D.*

The concrete application of Kas' theorem present two big difficulties:
1) how to write such a monodromy factorization?
2) how to verify that two factorizations are equivalent?
3) how to verify that two factorizations are not equivalent?

For the first question, we shall show how we can calculate such factorizations, using the reality principle: trying to write fibrations which are defined over $\mathbb{R}$, and possibly with all the critical values being real (if not, they come into conjugate pairs). In this case the paths we choose can be explicitly described. In particular, it helps to write the branch curves as smoothings of rather singular but explicitly given curves, union of graphs of rational functions.

The second question is quite hard, but, at least in the case of abc-surfaces, a big help came from a simple geometric idea.

In fact, consider our surfaces as bidouble covers $Z$

$$(***) \quad z_{a,b}^2 = f'_{2a,2b}(x,y)$$
$$w_{c,b}^2 = g'_{2c,2b}(x,y)$$

branched on a union of horizontal and of vertical lines: in this case the smooth surface $S$ is the blow up of $Z$ at the nodes lying over the nodes of $C' = \{f'(x,y) = 0\}$ and of $D' = \{g'(x,y) = 0\}$.

Changing the bidegree of $f'$ from $(2a, 2b)$ to $(2a+2, 2b)$, and the bidegree of $g'$ from $(2c, 2b)$ to $(2c-2, 2b)$ amounts geometrically to removing two vertical lines from the curve $D'$ and putting them back into the curve $C'$. Here we exploit the fortunate circumstance that $b = d$: we take the horizontal lines of $C'$ as the lines $y = j$, $j = 1, \dots 2b$, and the horizontal lines of $D'$ as the lines $y = -j$, $j = 1, \dots 2b$.

Consider then the manifold with boundary $M$, resolution of the branched cover of $\Delta \times \mathbb{P}^1$, where $\Delta$ is the unit disk in $\mathbb{C}$, described by

$$z^2 = \Pi_1^{2b}(y-i)$$
$$w^2 = (x^2 - \epsilon^2)\Pi_1^{2b}(y+i).$$

Then $S'$ is obtained from $S$ removing $M$ and attaching it back on the boundary of $\Delta$ after acting with the automorphism $\psi$ of order two such that $\psi(x,y,z,w) = (x, -y, w, z)$.

The result is that the factorizations for $S$ and $S'$ differ just in this beginning part, where the factors have been conjugated by $\psi$. Geometrically, one says that the fibrations of $S$ and $S'$ are obtained as two different fibre sums, i.e., one adds to the same factorization (the final part) a beginning factorization, once as it is, and another time conjugated by $\psi$. There is a very nice lemma, due to Auroux (mentioned also parenthetically by Kas), which gives a criterion for equivalence of these two beginning pieces of the respective factorizations:

**Lemma 3.4. (Auroux)** *Let $\tau$ be a Dehn twist and let $F$ be a factorization of a central element $\phi \in \mathfrak{Map}_g$, $\tau_1 \circ \tau_2 \circ \cdots \circ \tau_m = \phi$.*

*If there is a factorization $F'$ such that $F$ is Hurwitz equivalent to $\tau \circ F'$, then $(F)^\tau := \tau^{-1} F \tau$ is Hurwitz equivalent to $F$.*

*In particular, if $F$ is a factorization of the identity, $\psi = \Pi_h \tau'_h$, and $\forall h$ $\exists F'_h$ such that $F$ is Hurwitz equivalent to $\tau'_h \circ F'_h$, then the fibre sum with the Lefschetz pencil associated with $F$ yields the same Lefschetz pencil as the fibre sum twisted by $\psi$.*

*Proof.* If $\cong$ denotes Hurwitz equivalence, then

$$(F)^\tau \cong \tau \circ (F')^\tau \cong F' \circ \tau \cong (\tau)^{(F')^{-1}} \circ F' = \tau \circ F' \cong F.$$

*Q.E.D.*

**Corollary 3.5.** *Notation as above, assume that $F$, $\tau_1 \circ \tau_2 \circ \cdots \circ \tau_m = \phi$, is a factorization of the Identity and that $\psi$ is a product of the Dehn twists $\tau_i$ appearing in $F$. Then a fibre sum with the Lefschetz pencil associated with $F$ yields the same result as the same fibre sum twisted by $\psi$.*

*Proof.* We need only to verify that for each $h$ , there is $F'_h$ such that $F \cong \tau_h \circ F'_h$.

But this is immediately obtained by applying $h-1$ Hurwitz moves, the first one between $\tau_{h-1}$ and $\tau_h$, and proceeding further to the left till we obtain $\tau_h$ as first factor.

*Q.E.D.*

Hence in [CW04] the problem was reduced to proving that the isotopy class of $\psi$ was a product of certain Dehn twists appearing in the first part of the factorization.

Verifying isotopy of these diffeomorphisms was accomplished by constructing chains of loops in the complex fibre curve $C_0$, which lead to a dissection of $C_0$ into open cells. The problem was then solved choosing several associate Coxeter elements to express the given diffeomorphism $\psi$ as a product of those Dehn twists.

In the next section we shall instead concentrate on Question 3), which is related to our recent work.

## 4. How to distinguish factorizations up to Hurwitz equivalence, and stable equivalence

In this section the set up will be more algebraically oriented: the reason for doing so stems from the fact that in our applications we shall deal with two different type of groups, not only mapping class groups, but also braid groups, and moreover we shall have to deal, in the second case, with two different types of equivalence relations, the first one being the equivalence relation generated by simultaneous conjugation and by Hurwitz equivalence, the second one, introduced by Auroux and Katzarkov, is called stable equivalence and is more general.

We shall look carefully into the problem of distinguishing classes of factorizations, and for this reason we shall introduce a technical novelty allowing a new effective method for disproving stable-equivalence.

Assume that we have a group $G$. Then a factorization in $G$ of length $m$ is nothing else than an element in the Cartesian product $G^m$.

The product function associates to such an element $(\alpha_1, \dots \alpha_m) \in G^m$ its product $\alpha_1 \cdot \dots \cdot \alpha_m = a$.

The inverse image of the element $a$ under the product function is called the set of **factorizations of $a$ in $G$ of length $m$** and we write such a factorization of $a$ in $G$ with the notation

$$(*)\ \alpha_1 \circ \dots \circ \alpha_m = a.$$

There is a natural action of the group $Aut(G)$ on the set of length $m$ factorizations, in particular $G$ acts by conjugation, hence the Centralizer $Z(a)$ acts on the factorizations of $a$ by simultaneous conjugation,

$$(\alpha_1 \circ \dots \circ \alpha_m = a) \mapsto (\alpha_1^g \circ \dots \circ \alpha_m^g = a), \ \forall g \in Z(a),$$

where as before we set $a^g := g^{-1}ag$.

Hurwitz equivalence of factorizations is the equivalence generated by Hurwitz moves, where the i-th Hurwitz move leaves all the $\alpha_j$'s with $j \neq i, i+1$ unchanged, while it replaces $\alpha_{i+1}$ with $\alpha_i$, and $\alpha_i$ with $\alpha_i\alpha_{i+1}\alpha_i^{-1}$, so that the product of the factorization remain unchanged.

Hurwitz equivalence classes are nothing else than orbits for the action of the braid group $\mathcal{B}_m$ on the set of factorizations of a given element $a$.

Recall (cf. e.g. [Bir74]):

**Theorem 4.1.** *The braid group $\mathcal{B}_n$ is the subgroup of $\mathfrak{Map}(\mathbb{C} - \{1, \dots n\})$ given by the diffeomorphisms which are the identity outside of a circle with centre the origin and radius $2n$. It has the Artin presentation*

$$\langle \sigma_1, \dots \sigma_{n-1} | \sigma_i\sigma_j = \sigma_j\sigma_i \text{ for } |i-j| \geq 2,\ \sigma_i\sigma_{i+1}\sigma_i = \sigma_{i+1}\sigma_i\sigma_{i+1}\ , \forall i \leq n-2 \rangle.$$

Through this isomorphism the standard Artin generators $\sigma_j$ of $\mathcal{B}_n$ ( $j = 1, \dots n-1$) correspond to standard half-twists: $\sigma_j$ is the diffeomorphism of $(\mathbb{C}, \{1, \dots n\})$ isotopic to the homeomorphism of $(\mathbb{C}, \{1, \dots n\})$ which is a rotation of 180 degrees on the disc with centre $j + 1/2$ and radius $1/2$ , on a circle with the same centre and radius $1/2 + t/4$ is the identity for $t \geq 1$, and a rotation of angle $180(1-t)$ degrees if $0 \leq t \leq 1$.

Since $\mathcal{B}_n$ is a subgroup of $\mathfrak{Map}(\mathbb{C} - \{1, \dots n\})$ we have an obvious action of $\mathcal{B}_n$ on the free group $\pi_1(\mathbb{C} - \{1, \dots n\})$. Taking as base point the complex number $p := -2ni$, one has a geometric basis $\gamma_1, \dots \gamma_n$ on which $\mathcal{B}_n$ acts through the

**Hurwitz action of the braid group**

- $\sigma_i(\gamma_i) = \gamma_{i+1}$
- $\sigma_i(\gamma_i\gamma_{i+1}) = \gamma_i\gamma_{i+1}$ hence $\sigma_i(\gamma_{i+1}) = \gamma_{i+1}^{-1}\gamma_i\gamma_{i+1}$
- $\sigma_i(\gamma_j) = \gamma_j$ for $j \neq i, i+1$.

The Hurwitz action leaves the product $\gamma_1\gamma_2 \dots \gamma_n$ invariant.

We can summarize the above discussion in the following

**Definition 4.2.** *Let $G$ be a group. Consider the partition of its Cartesian product $G^n$ given by the product map, which to an ordered n-tuple $(g_1, g_2, \dots g_n)$ assigns the product $g := g_1 \cdot g_2 \cdots g_n \in G$.*

*The subsets of the partition of $G^n$ are called the* **factorizations of an element** $g \in G$. *The orbits of the action of $\mathcal{B}_n$, which preserves the partition of $G^n$, are called* **Hurwitz equivalence classes of factorizations**.

*Two factorizations are said to be* **M-equivalent** *(this means: equivalent by Hurwitz equivalence and simultaneous conjugation, and the M- stands for monodromy) if they are in the same $\mathcal{B}_n \times G$-orbit, where $G$ acts by (we use the notation previously introduced $a^b := b^{-1}ab$) the transformation*

$$Int_b \colon (g_1 \circ g_2 \circ \cdots \circ g_n \mapsto (g_1)^b \circ (g_2)^b \circ \cdots \circ (g_n)^b.$$

**Remark 4.3.** *Given a group homomorphism $\phi : G \to G'$,and a factorization in $G$, $g_1 \circ g_2 \circ \cdots \circ g_n = a$, we obtain a corresponding factorization*

$$\phi(g_1) \circ \phi(g_2) \circ \cdots \circ \phi(g_n)) = \phi(a),$$

*through the application of $\phi^n$.*

*$\phi^n$ preserves Hurwitz equivalence, and simultaneous conjugation. Therefore, a basic method to disprove equivalence of factorizations is to disprove equivalence of homomorphic factorizations in a simpler group $G'$.*

*This is essentially the main underlying idea.*

There are some obvious invariants for a factorization class:

**Proposition 4.4.** *Consider the set of Hurwitz equivalence classes of factorizations of a fixed element $a \in G$:*

$$(*)\ \alpha_1 \circ \cdots \circ \alpha_m = a.$$

*Then the basic invariants of a Hurwitz factorization class are:*

*i) the subgroup $H$ generated by the elements $\alpha_1, \ldots \alpha_m$*

*ii) the function, defined on the set $\mathcal{C}_H$ of $H$-conjugacy classes in $H$, assigning to each conjugacy class $A \in \mathcal{C}_H$ the integer $\sigma(A) \in \mathbb{N}$ counting how many of the elements $\alpha_1, \ldots \alpha_m$ belong to $A$.*

*Consider instead the set of M-equivalence classes of factorizations of a fixed element $a \in G$:*

$$(*)\ \alpha_1 \circ \cdots \circ \alpha_m = a.$$

*Then the basic invariants of an M-factorization class are, $Z(a)$ being the centralizer of the element $a \in G$:*

*1) the $Z(a)$-conjugacy class of the subgroup $H$ generated by the elements $\alpha_1, \ldots \alpha_m$, in particular the isomorphism class of $H$ as an abstract group.*

*2) the $Z(a)$ class of the function $\sigma \colon \mathcal{C}_H \to \mathbb{N}$, in particular, the cardinalities $\sigma^{-1}(n)$, for $n \in \mathbb{N}$.*

In the holomorphic case the above invariants are well defined; however, in the symplectic world Auroux and Katzarkov had to introduce a broader equivalence of factorizations, where the length of the factorization is not fixed a priori. They called it $m$-equivalence, but we prefer to call it stable-equivalence, or S-equivalence.

**Definition 4.5.** *Consider a group $G$, an element $a \in G$, and a set $\mathcal{B}$ of elements $\beta_j \in G$.*

*We define* **stable**-*equivalence with respect to $\mathcal{B}$ simply by considering the equivalence relation generated by Hurwitz equivalence, simultaneous conjugation and creation/cancellation of consecutive factors $\beta_j \circ \beta_j^{-1}$.*

Then **refined invariants** of the stable equivalence class of a factorization are obtained as follows:

I) let $H$ be the subgroup of $G$ generated by the $\alpha_i$'s and let $\hat{H}$ be the subgroup of $G$ generated by the $\alpha_i$'s and by the $\beta_j$'s (in our case, $H$ will be called the monodromy group, and $\hat{H}$ the stabilized monodromy group).

Then the $Z(a)$-conjugacy class of $\hat{H}$ is a first invariant, in particular the isomorphism class of $\hat{H}$.

II) Let $\mathcal{C}_H$ be the set of $H$-conjugacy classes $A$ in the group $H$, and let $\hat{\mathcal{C}}$ be the set of $\hat{H}$-conjugacy classes in the group $\hat{H}$, so that we have a natural map $\mathcal{C} \to \hat{\mathcal{C}}$, with $A \mapsto \hat{A}$.

Write $\hat{\mathcal{C}}$ as a disjoint union $\hat{\mathcal{C}}^+ \cup \hat{\mathcal{C}}^- \cup \hat{\mathcal{C}}^0$, where $\hat{\mathcal{C}}^0$ is the set of conjugacy classes of elements $a$ which are conjugate to their inverse $a^{-1}$, and where $\hat{\mathcal{C}}^-$ is the set of the inverse conjugacy classes of the classes in $\hat{\mathcal{C}}^+$.

Associate now to the factorization $\alpha_1 \circ \cdots \circ \alpha_m = 1$ the function $s : \hat{\mathcal{C}}^+ \to \mathbb{Z}$ such that $s(c)$, for $c \in \hat{\mathcal{C}}^+$, is the algebraic number of occurrences of $c$ in the sequence of conjugacy classes of the $\alpha_j$'s (i.e., an occurrence of $c^{-1}$ counts as $-1$ for $s(c)$).

The $Z(a)$ class of the function $s : \hat{\mathcal{C}}^+ \to \mathbb{Z}$ is our second and most important invariant. From it one can derive an easier invariant, i.e., the $Z(a)$ class of the function induced by the absolute value $|s|$ on the set $\hat{\mathcal{C}}^{+-}$ of pairs of distinct conjugacy classes, one inverse of the other. In particular, the values $|s|^{-1}(n)$ for $n \in \mathbb{N}$.

**Remark 4.6.** *The calculation of the function $s : \hat{\mathcal{C}}^+ \to \mathbb{Z}$ presupposes however a detailed knowledge of the group $\hat{H}$. For this reason in our paper* [CLW] *we resorted to using some coarser derived invariant.*

**Substrategy I.**

Assume that we can write $\{\alpha_1, \ldots, \alpha_m\}$ as a disjoint union $\mathcal{A}_1 \cup \mathcal{A}_2 \cup \mathcal{D} \cup \mathcal{A}'_1 \cup \mathcal{A}'_2$, such that the set $\mathcal{A}_j$, $j = 1, 2$, is contained in a conjugacy class $A_j \subset H$, the set $\mathcal{A}'_j$, $j = 1, 2$, is contained in the conjugacy class $A_j^{-1} \subset H$. Assume that the elements in $\mathcal{A}_1 \cup \mathcal{A}_2$ are contained in a conjugacy class $C$ in $G$ such that $C \cap C^{-1} = \emptyset$ but that the set $\mathcal{D}$ is disjoint from the union of the two conjugacy classes of $G$, $C \cup C^{-1}$.

If we then prove that $\hat{A}_1 \neq \hat{A}_2$ (this of course implies $A_1 \neq A_2$) we may assume without loss of generality that $\hat{A}_1, \hat{A}_2 \in \hat{\mathcal{C}}^+$, and then the unordered pair of positive numbers $(|s(\hat{A}_1)|, |s(\hat{A}_2|)$ is our derived numerical invariant of the factorization (and can easily be calculated from the cardinalities of the four sets as $(||\mathcal{A}_1| - |\mathcal{A}'_1||, ||\mathcal{A}_2| - |\mathcal{A}'_2||)$.

**Substrategy II.**

This is the strategy to show that $\hat{A}_1 \neq \hat{A}_2$ and goes as follows.

Assume further that we have another subgroup of $G$, $\tilde{H} \supset \hat{H}$, and a group homomorphism $\rho : \tilde{H} \to \Sigma$ such that $\rho(\mathcal{B}) = 1$. Assume also that the following key property holds.

**Key property.**

For each element $\alpha_j \in A_j \subset H$ there exists an element $\tilde{\alpha}_j \in \tilde{H}$ such that

$$\alpha_j = \tilde{\alpha}_j^2,$$

and moreover that this element is unique in $\tilde{H}$ (a fortiori, it will suffice that it is unique in $G$).

Proving that $\rho(\tilde{\alpha}_1)$ is not conjugate to $\rho(\tilde{\alpha}_2)$ under the action of $\rho(H) = \rho(\hat{H}) \subset \Sigma$ shows finally that $\alpha_1$ is not conjugate to $\alpha_2$ in $\hat{H}$.

Since, if there is $h \in \hat{H}$ such that $\alpha_1 = h^{-1}\alpha_2 h$, then $\tilde{\alpha}_1^2 = \alpha_1 = (h^{-1}\tilde{\alpha}_2 h)^2$, whence $\tilde{\alpha}_1 = h^{-1}\tilde{\alpha}_2 h$, a contradiction.

In our concrete case, we are able to determine the braid monodromy group $H$, and we observe that, since we have a so called cuspidal factorization, all the factors $\alpha_i$

belong to only four conjugacy classes in the group $G$ (the classes of $\sigma_1$,$\sigma_1^2$, $\sigma_1^{-2}$,$\sigma_1^3$ in the braid group), corresponding geometrically to vertical tangencies of the branch curve, respectively positive nodes, negative nodes and cusps. Three of these classes are positive and only one is negative (the one of $\sigma_1^{-2}$).

We see in [CLW] that the nodes belong to conjugacy classes $A_1, A_2, A_1^{-1}, A_2^{-1}$ in $H$, the positive nodes belonging to $A_1 \cup A_2$, and we show, using a representation $\rho$ of a certain subgroup $\tilde{H}$ of 'liftable' braids (i.e., braids which centralize the monodromy homomorphism, whence are liftable to the mapping class group of a curve $D_0$ associated to the monodromy homomorphism) into a symplectic group $\Sigma$ with $\mathbb{Z}/2$ coefficients, that the classes $\hat{A}_1, \hat{A}_2$ are distinct in $\hat{H}$. Moreover, these are classes in $\mathcal{C}^+$, since these are positive classes in the braid group. We calculate then easily the above function for these two conjugacy classes, i.e., the pair of numbers $(s(\hat{A}_1), s(\hat{A}_2))$.

## 5. Bidouble covers of the quadric and their symplectic perturbations.

In order to see in more detail how the algebraic considerations of the previous sections apply to our situation, let us go back to our equations

$$(***) \qquad z_{a,b}^2 = f_{2a,2b}(x,y) + w_{c,d}\Phi_{2a-c,2b-d}(x,y)$$

$$w_{c,d}^2 = g_{2c,2d}(x,y) + z_{a,b}\Psi_{2c-a,2d-b}(x,y)$$

where $f, g$ are bihomogeneous polynomials as before, and instead we shall allow $\Phi, \Psi$, in the case where for instance the degree relative to $x$ is negative, to be antiholomorphic. In other words, we allow $\Phi, \Psi$ to be sections of certain line bundles which are dianalytic (holomorphic or antiholomorphic) in each variable $x, y$.

We can more generally consider the direct sum $\mathbb{V}$ of two complex line bundles $\mathbb{L}_1 \oplus \mathbb{L}_2$ on a compact complex manifold $X$, and the subset $Z$ of $\mathbb{V}$ defined by the following pair of equations,

$$z^2 = f(x) + w\ \Phi(x)$$
$$w^2 = g(x) + z\ \Psi(x)$$

where $f, g$ are respective holomorphic sections of the line bundles $\mathbb{L}_1^{\otimes 2}, \mathbb{L}_2^{\otimes 2}$ (we shall also write $f \in H^0(\mathcal{O}_X(2L_1)), g \in H^0(\mathcal{O}_X(2L_2))$, denoting by $L_1, L_2$ the associated Cartier divisors), and where $\Phi$ is a differentiable section of $\mathbb{L}_1^{\otimes 2} \otimes \mathbb{L}_2^{\otimes -1}$, $\Psi$ is a differentiable section of $\mathbb{L}_2^{\otimes 2} \otimes \mathbb{L}_1^{\otimes -1}$.

The key observation is that, for $\Phi \equiv \Psi \equiv 0$, then $\pi : Z \to X$ is a Galois cover with group $(\mathbb{Z}/2\mathbb{Z})^2$, while, if $\Phi, \Psi$ are not $\equiv 0$, then we obtain a covering whose monodromy group is the symmetric group $\mathfrak{S}_4$.

We have the following Lemmas from [CLW].

**Lemma 5.1.** *Assume that the two divisors* $\{f = 0\}$ *and* $\{g = 0\}$ *are smooth and intersect transversally. Then, for* $|\Phi| << 1$, $|\Psi| << 1$, $Z$ *is a smooth submanifold of* $\mathbb{V}$, *and the projection* $\pi : \mathbb{V} \to X$ *induces a finite covering* $Z \to X$ *of degree* 4, *with ramification divisor (i.e., critical set)* $R := \{4zw = \Phi\Psi\}$, *and with branch divisor (i.e., set of critical values)* $\Delta = \{\delta(x) = 0\}$, *where*

$$-\frac{1}{16}\delta = -f^2g^2 - \frac{9}{8}fg(\Phi\Psi)^2 + (\Psi)^2f^3 + (\Phi)^2g^3 + \frac{27}{16^2}(\Phi\Psi)^4.$$

We want to analyse now the singularities that $\Delta$ has, for a general choice of $\Phi, \Psi$. The singularities which are easily spotted deserve a special definition.

**Definition 5.2.** *A point* $x$ *is said to be a* **trivial singularity** *of* $\Delta$ *if it is either a point where* $f = \Phi = 0$, *or a point where* $g = \Psi = 0$.

*Z is said to be* **mildly general** *if*

*(*) at any point where* $f = \Phi = 0$, *we have* $g \cdot \Psi \neq 0$, *and symmetrically at any point where* $g = \Psi = 0$, *we have* $f \cdot \Phi \neq 0$

Besides the trivial singularities, we show in general that the other singularities are located in a neighbourhood of the points with$f(x) = g(x) = 0$.

**Lemma 5.3.** *If* $Z$ *is mildly general, and* $|\Phi| << 1$, $|\Psi| << 1$, *the singular points of* $\Delta$ *which are not the trivial singularities of* $\Delta$ *occur only for points* $p \in Z_x$ *(*$Z_x$ *being the fibre over the point* $x$*) of multiplicity* 3 *which lie over arbitrarily small neighbourhoods of the points* $x'$ *with* $f(x') = g(x') = 0$.

We can describe more precisely the singularities of $\Delta$ in the case where $X$ is a complex surface.

**Proposition 5.4.** *Assume that* $X$ *is a compact complex surface, that the two divisors* $\{f = 0\}$ *and* $\{g = 0\}$ *are smooth and intersect transversally in a set* $M$ *of* $m$ *points. Then, for* $|\Phi| << 1$, $|\Psi| << 1$, *and for* $Z$ *mildly general,* $Z$ *is a smooth submanifold of* $\mathbb{V}$, *and the projection* $\pi : \mathbb{V} \to X$ *induces a finite covering* $Z \to X$ *of degree* 4, *with smooth orientable ramification divisor* $R := \{4zw = \Phi\Psi\}$, *and with branch divisor* $\Delta$ *having as singularities precisely*

*1)* $3m$ *cusps lying (in triples) in an arbitrarily small neighbourhood of* $M$, *and moreover*

*2) the trivial singularities, which are nodes if the curve* $\{f = 0\}$ *intersects transversally* $\{\Phi = 0\}$, *respectively if the curve* $\{g = 0\}$ *intersects transversally* $\{\Psi = 0\}$.

Since $R$ is orientable, it follows that if the trivial singularities of $\Delta$ are nodes, then the intersection number of the two branches is $\pm 1$; the local selfintersection number is exactly equal to $+1$ when the two orientations of the two branches combine to yield the natural (complex) orientation of $X$.

We can calculate the precise number of singularities, and of the positive, respectively negative nodes of $\Delta$, in the special case we are interested in, namely, of a perturbed bidouble cover of the quadric.

**Definition 5.5.** *Given four integers* $a, b, c, d \in \mathbb{N}_{\geq 3}$, *the so called* **dianalytic perturbations of simple bidouble covers** *are the 4-manifolds defined by equations*

$$z^2 = f_{(2a,2b)}(x_0, x_1; y_0, y_1) + w \ \Phi_{(2a-c,2b-d)}(x_0, x_1; y_0, y_1)$$
$$w^2 = g_{(2c,2d)}(x_0, x_1; y_0, y_1) + z \ \Psi_{(2c-a,2d-b)}(x_0, x_1; y_0, y_1)$$

*where* $f, g$ *are bihomogeneous polynomials of respective bidegrees* $(2a, 2b)$, $(2c, 2d)$, *and where* $\Phi, \Psi$ *are polynomials in the ring*

$$\mathbb{C}[x_0, x_1, y_0, y_1, \bar{x_0}, \bar{x_1}, \bar{y_0}, \bar{y_1}],$$

*bihomogeneous of respective bidegrees* $(2a - c, 2b - d)$, $(2c - a, 2d - b)$ *(the ring is bigraded here by setting the degree of* $\bar{x_i}$ *equal to* $(-1, 0)$ *and the degree of* $\bar{y_i}$ *equal to* $(0, -1)$ *).*

*We choose moreover* $\Phi, \Psi$ *to belong to the subspace where all monomials are either separately holomorphic or antiholomorphic, i.e., they admit no factor of the form* $x_i\bar{x_j}$ *or of the form* $y_i\bar{y_j}$.

**Remark 5.6.** *Indeed it is convenient to pick* $\Phi$ *(respectively,* $\Psi$*) to be a product*

$$\Phi = \Phi_1(x)\Phi_2(y),$$

*where* $\Phi_1$ *is a product of linear forms, either all holomorphic or all antiholomorphic, and similarly for* $\Phi_2$ *(respectively, for* $\Psi_1$, $\Psi_2$*).*

As explained in [Cat08](pages 134-135) to an antiholomorphic homogeneous polynomial $P(\bar{x_0}, \bar{x_1})$ of degree $m$ we associate a differentiable section $p$ of the tensor power $\mathbb{L}^{\otimes m}$ of the tautological (negative) subbundle, such that $p(x) := P(1, \bar{x})$ inside a big disc $B(0, r)$ in the complex line having centre at the origin and radius $r$.

We make the following assumptions on the polynomials $f, g, \Phi, \Psi$:

(1) The algebraic curves $C := \{f = 0\}$ and $D := \{g = 0\}$ are smooth
(2) $C$ and $D$ intersect transversally at a finite set $M$ contained in the product $B(0, r)^2$ of two big discs in $\mathbb{C}$, and at these points both curves have non vertical tangents
(3) for both curves $C$ and $D$ the first projection on $\mathbb{P}^1$ is a simple covering and moreover the vertical tangents of $C$ and $D$ are all distinct
(4) the associated perturbation is sufficiently small and mildly general
(5) the trivial singularities are nodes with non vertical tangents, contained in $B(0, r)^2$
(6) the cusps of $\Delta$ have non vertical tangent.

Considering then the branch curve $\Delta$ of the perturbed bidouble cover, the defining equation $\delta$ is bihomogeneous of bidegree $(4(a + c), 4(b + d))$.

We have that $\Delta$ has exactly $k := 12(ad + bc)$ cusps, coming from the $m := 4(ad + bc)$ points of the set $M = C \cap D$.

If $\Phi, \Psi$ are holomorphic, then $\Delta$ has exactly

$$\nu := 4(2ab + 2cd - ad - bc)$$

(positive) nodes.

In general, the above number $\nu$ equals the difference $\nu^+ - \nu^-$ between the number of positive and the number of negative nodes.

In fact the nodes of $\Delta$ occur only for the trivial singularities.

A trivial singularity $f = \Phi = 0$ yields a node with the same tangent cone as

$$f^2 - \Phi^2 \cdot g(x_0).$$

By remark 5.6 we may assume that there are local holomorphic coordinates $f, \hat{\Phi}$ such that either $\Phi =$ unit $\cdot \hat{\Phi}$ or $\Phi =$ unit $\cdot \overline{\hat{\Phi}}$.

In the first case we get the tangent cone of a holomorphic node, hence a **positive** node, in the second case we get the tangent cone of an antiholomorphic node, hence a **negative** node.

We summarize the results in the following

**Lemma 5.7.** *The number $t_f$ of vertical tangents for the curve $C$ is $t_f = 4(2ab - a)$, while the number $t_g$ of vertical tangents for the curve $D$ is $t_g = 4(2cd - c)$. For a general choice of $\Phi, \Psi$, the number of vertical tangents $t$ of $\Delta$ equals $t = 2t_f + 2t_g + m$, where $m := 4(ad + bc)$. The genus of the Riemann surface $R$ equals $g(R) = 1 + 16(a+c)(b+d) - 4(a+b+c+d) - k - \nu$, where $k := 3m = 12(ad + bc)$ is the number of cusps of $\Delta$ and $\nu := \nu^+ - \nu^-$ is the number of nodes of $\Delta$, counted with sign.*

**Remark 5.8.** *1) Observe that the fact that $p : R \to \mathbb{P}^1_{\mathbb{C}}$ has a finite number of critical points implies immediately that $p$ is finite, since there is then a finite set in $\mathbb{P}^1_{\mathbb{C}}$ such that over its complement we have a covering space. Thus $p$ is orientation preserving and each non critical point contributes positively to the degree of the map $p$. Hence some of the above calculations are obtained from Hurwitz' ramification formula.*

*2) Consider now a real analytic 1-parameter family $Z(\eta)$ such that the perturbation terms $\Phi_\eta, \Psi_\eta \to 0$. Then we have a family of ramification points $P_i(\eta)$ which tend to the ramification points of $p_0 : R_0 \to \mathbb{P}^1_{\mathbb{C}}$. For $R_0$, a double cover of $C \cup D$, we have $2t_f$ ramification points lying over the $t_f$ vertical tangents of $C$, $2t_g$ ramification points lying over the $t_g$ vertical tangents of $D$, whereas the $m$ nodes of $R_0$ are limits of 4 ramification points, three cusps and a simple vertical tangent, cf. the analysis made in* [CW05]. *This is the geometric reason why $t = 2t_f + 2t_g + m$; it also tells us where to look for the vertical tangents of $\Delta$.*

In practice, in order to effectively compute the braid monodromies, it is convenient to view the curves $C$ and $D$ as small real deformations of reducible real curves with only nodes (with real tangents) as singularities, and to consider real polynomials $\Phi, \Psi$, such that the trivial singularities are also given by real points. By the results of [CW05], from the perturbation of each proper node of $C \cup D$, i.e., of a point of $M = C \cap D$, we shall obtain a real vertical tangent, a real cusp, and two immaginary cusps.

We begin with real polynomials $f^\times$ and $g^\times$ defining nodal curves $C^\times$, respectively $D^\times$.

For the sake of simplicity we replace here bihomogeneous polynomials $f(x_0, x_1; y_0, y_1)$ by their restrictions to the affine open set $x_0 = 1, y_0 = 1$ and write $f(x, y)$ for $f(1, x; 1, y)$.

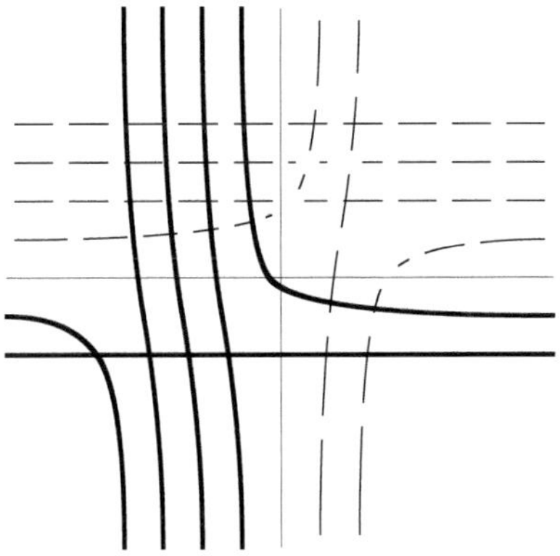

FIGURE 1. Real part of $\Delta_f$ and $\Delta_g$ for $(a,b,c,d) = (1,2,2,1)$

We let then $f^\times := f_1 \cdots f_{2b}$ defining $f_i := y - 2i$ for i=2, …, $2b$ but setting

$$f_1 := (y-2)\prod_{i=1}^{2a}(x-2i) + \prod_{i=1}^{2a-1}(x-2i-1),$$

and we define similarly $g^\times := g_1 \cdots g_{2d}$ setting $g_i = y + 2i$ for i=2, …, $2d$ except that we set

$$g_1 := (y+2)\prod_{i=1}^{2c}(x+2i) - \prod_{i=1}^{2c-1}(x+2i+1).$$

**Remark 5.9.** *Note that the equations of $f_1$ and $g_1$ are chosen in such a way that their zero sets are graphs of rational functions $\tilde{f}_1$, respectively $\tilde{g}_1$, of $x$ which, regarded as maps from $\mathbb{C}$ to $\mathbb{C}$, preserve the real line. Moreover, $\tilde{f}_1$ preserves both the upper and lower halfplane, while $\tilde{g}_1$ exchanges them.*

*In fact, $\tilde{f}_1$ has no critical points on $\mathbb{P}^1_{\mathbb{R}}$ and*

$$deg(\tilde{f}_1|_{\mathbb{P}^1_{\mathbb{R}}})\colon \mathbb{P}^1_{\mathbb{R}} \to \mathbb{P}^1_{\mathbb{R}} = 2a = deg(\tilde{f}_1).$$

*Similarly for $\tilde{g}_1$.*

Given $f^\times$, consider the polynomial $f : f^\times + c_f$, where $c_f$ is a small constant; likewise consider $g : g^\times + c_g$. Adding these small constants to the respective equations of $C^\times$ and $D^\times$ we get polynomials $f$ and $g$ which define smooth curves $C$ and $D$. We have more precisely:

**Proposition 5.10.** *If the constants $c_f$, $c_g$, $\eta$ are chosen sufficiently small, the polynomials $f, g, \eta\Phi, \eta\Psi$ thus obtained satisfy the following list of hypotheses (cf. p. 75), for $r > 8(a+b+c+d)$:*

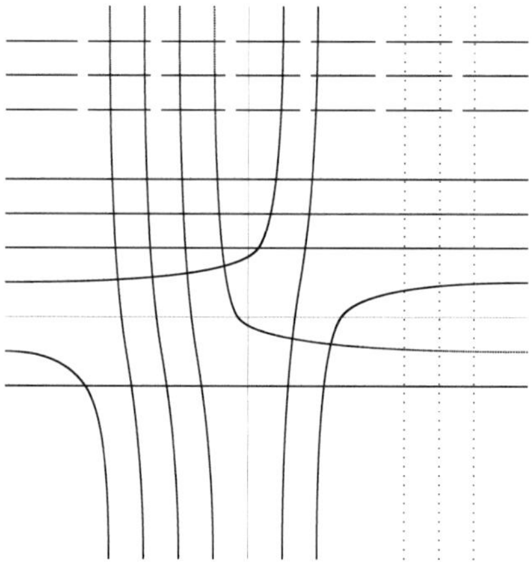

FIGURE 2. Zero sets of $\Phi$ (dashed) and $\Psi$ (dotted) added

(1) *The algebraic curves $C = \{f = 0\}$ and $D = \{g = 0\}$ are smooth*
(2) *$C$ and $D$ intersect transversally at a finite set $M$ contained in the product $B(0, r)^2$ and at these points both curves have non vertical tangents*
(3) *for both curves $C$ and $D$ the first projection on $\mathbb{P}^1$ is a simple covering and moreover the vertical tangents of $C$ and $D$ are all distinct*
(4) *the associated perturbation is mildly general*
(5) *the trivial singularities of $\Delta$ are nodes with non vertical tangents, contained in $B(0, r)^2$*
(6) *the cusps of $\Delta$ have non vertical tangent*

## 6. Bidouble covers of the quadric and their Braid Monodromy factorization types.

In the previous section we showed how to obtain a symplectic 4-manifold which (we call a dianalytic perturbation and) is canonically symplectomorphic to the bidouble cover $S$ we started with.

Now we have that, for general choice of $f, g, \Phi, \Psi$, the projection $\pi$ onto $Q = \mathbb{P}^1 \times \mathbb{P}^1$ is **generic** and its branch curve $\Delta$ (the locus of the critical values) is a dianalytic curve with nodes and cusps as only singularities. In this way we may however have introduced also negative nodes, i.e., nodes which in local holomorphic coordinates are defined by the equation

$$(y - \bar{x})(y + \bar{x}) = 0.$$

Now, the projection onto the first factor $\mathbb{P}^1$ gives two monodromy factorizations:

**Definition 6.1.** *The* **vertical braid monodromy factorization** *of a bidouble cover is the factorization of the identity in the braid group $\mathrm{Br}_n$ of the sphere $\mathbb{P}^1$ associated to the* *first projection onto* $\mathbb{P}^1$ *of the pair* $(Q, \Delta)$. *Here* $n = 4(b+d)$ *is the vertical degree of* $\Delta$, *and the monodromy factorization describes the movement of* $n = 4(b+d)$ *points in a fibre* $F_0 = \mathbb{P}^1$.

The Braid Monodromy factorization has factors in the braid group of the sphere $\mathrm{Br}_n$:

$$\mathrm{Br}_n \quad = \quad \left\langle \sigma_1, \ldots, \sigma_{n-1} \;\middle|\; \begin{array}{l} \sigma_i\sigma_j = \sigma_j\sigma_i, \quad \text{if } |i-j| > 1 \\ \sigma_i\sigma_{i+1}\sigma_i = \sigma_{i+1}\sigma_i\sigma_{i+1} \\ \sigma_1 \cdots \sigma_{n-1}\sigma_{n-1} \cdots \sigma_1 = 1 \end{array} \right\rangle$$

The factors are of three following types

(1) each cusp of $\Delta$ (which has non vertical tangent) yields a factor which is a conjugate of $\sigma_1^3$
(2) each vertical tangency (this takes place in smooth points only) yields a factor which is a conjugate of $\sigma_1$
(3) each trivial singularity yields a factor which is a conjugate of $\sigma_1^2$ in the case where we have a positive node, and a factor which is a conjugate of $\sigma_1^{-2}$ in the case where we have a negative node.

We already know, by Kas' theorem, that the differentiable type of the Lefschetz fibration $F : S \to \mathbb{P}^1$ given by the composition of the perturbed covering $\pi : S \to Q$ with the first projection onto $\mathbb{P}^1$ is encoded in the M-equivalence class of the factorization of the identity in the Mapping class group $\mathcal{M}ap_g$ corresponding to the monodromy factorization of this Lefschetz fibration (here $g$ is the genus of the base fibre $C_0$ of $F$) .

The second factorization is the homomorphic image of the Braid Monodromy factorization.

It is obtained since the braid monodromy factors are contained in a subgroup of the Braid group $\mathrm{Br}_n$, namely, the subgroup of liftable braids, corresponding to diffeomorphisms which lift to the fibre $C_0$ of the Lefschetz fibration $F : S \to \mathbb{P}^1$ which lies over $F_0$.

More precisely, we have the covering $C_0 \backslash F^{-1}(\Delta) \to F_0 \backslash \Delta$, and let $\mu : \pi_1(F_0 \backslash \Delta) \to \mathfrak{S}_4$ be its monodromy homomorphism: then the braid monodromy factors are contained in the subgroup

$$\mathcal{L} := \{\sigma \in \mathrm{Br}_n \,|\mu\sigma_* = \mu\}.$$

And each braid in $\mathcal{L}$ determines a diffeomorphism of the pair $(C_0, C_0 \cap \Delta)$ which in turn gives us an element in $\mathcal{M}ap_g$ .

The easy but important observation is that the homomorphism

$$\mathcal{L} \to \mathcal{M}ap_g$$

is far from being injective.

In fact, the factors of the Braid Monodromy factorization map to:

(1) each factor corresponding to a cusp of $\Delta$ maps to the identity
(2) each factor coming from a vertical tangency maps to a Dehn twist which is the lift of a conjugate of $\sigma_1$

(3) each factor coming from a trivial singularity , be it a positive or a negative node, maps to the identity.

Before we state our characterization of the Braid Monodromy factorization, it will be convenient to briefly describe the geometric picture of the degree 4 covering $F : C_0 \to F_0$.

In the case where $b = d$ we may start with an extremely symmetric picture, taking the bidouble cover $F'_0$ of $\mathbb{P}^1_{\mathbb{C}}$ of equation

- $z^2 = f(y), w^2 = g(y) := f(-y)$
  and of the automorphism $\psi$ given by
- $\psi(y, z, w) := (-y, w, z)$

and we see immediately that the curve $C_0$ has an automorphism group generated by the previously mentioned automorphism $\psi$ and by the Galois group of the covering, generated by

$$g_1(y, z, w) := (y, -z, w), g_2(y, z, w) := (y, z, -w).$$

This group is indeed generated by $g_1$ and the order 4 element $g_4(y, z, w) := (-y, w, -z)$; since $g_1 g_4 g_1 g_4 = 1$, it is the dihedral group $D_4$.

We recall here once more the role of the automorphism $\psi$, namely that the following was proven in [CW04] :

**Proposition 6.2.** *The monodromy of $S$ over the unit circle $\{x||x| = 1\}$ is trivial, and the pair $(C_0, \psi)$ yields the fibre over $x = 1$, considered as a differentiable 2-manifold, together with the isotopy class of the map yielding the twisted fibre sum corresponding to $S'$.*

In the general case, perturbing the equations of $C_0$ amounts to making the degree 4 morphism generic. In this context we can view the effect of the perturbation as replacing the branch locus

$$B'_0 := \{f^{\times}(y) = 0\} \cup \{g^{\times}(y) = 0\} = \{1, \dots, 2b\} \cup \{-1, \dots, -2d\}$$

with a new 'doubled' branch locus

$$B_0 := \{f(y)g(y) = 0\} = \{1 \pm \epsilon\sqrt{-1}, \dots, 2b \pm \epsilon\sqrt{-1}\} \cup \{-1 \pm \epsilon\sqrt{-1}, \dots, -2d \pm \epsilon\sqrt{-1}\}.$$

Before giving the full vertical Braid monodromy factorization, we show a simpler byproduct of our description of the Braid monodromy: it makes effective and precise the contents of the above intuitive picture.

The theorem is stated in its most understandable form, namely a pictorial one, recalling that any segment in the plane determines a (non standard) half twist in $\mathrm{Br}_n$, $n = 4(b+d)$.

**Theorem 6.3.** *There is a braid monodromy factorization of the curve $\Delta$ associated to a bidouble cover $S$ of type $a, b, c, d$ whose braid monodromy group $H \subset \mathrm{Br}_{4(b+d)}$ is generated, unless we are in the cases*

$$\text{(I) } c = 2a \text{ and } d = 2b \text{ , or (II) } a = 2c \text{ and } b = 2d$$

*by the following powers of half-twists:* $\sigma_{a_i}, \sigma_{c_i}$ *for* $i = 1, \dots 2b - 1$, $\sigma_{b_\imath}, \sigma_{d_\imath}$, *for* $\imath = 1, \dots, 2d - 1$, $\sigma^2_{p_{2b}}, \sigma^2_{q_{2d}}, \sigma_s, \sigma^3_{u'}, \sigma^3_{u''}$.

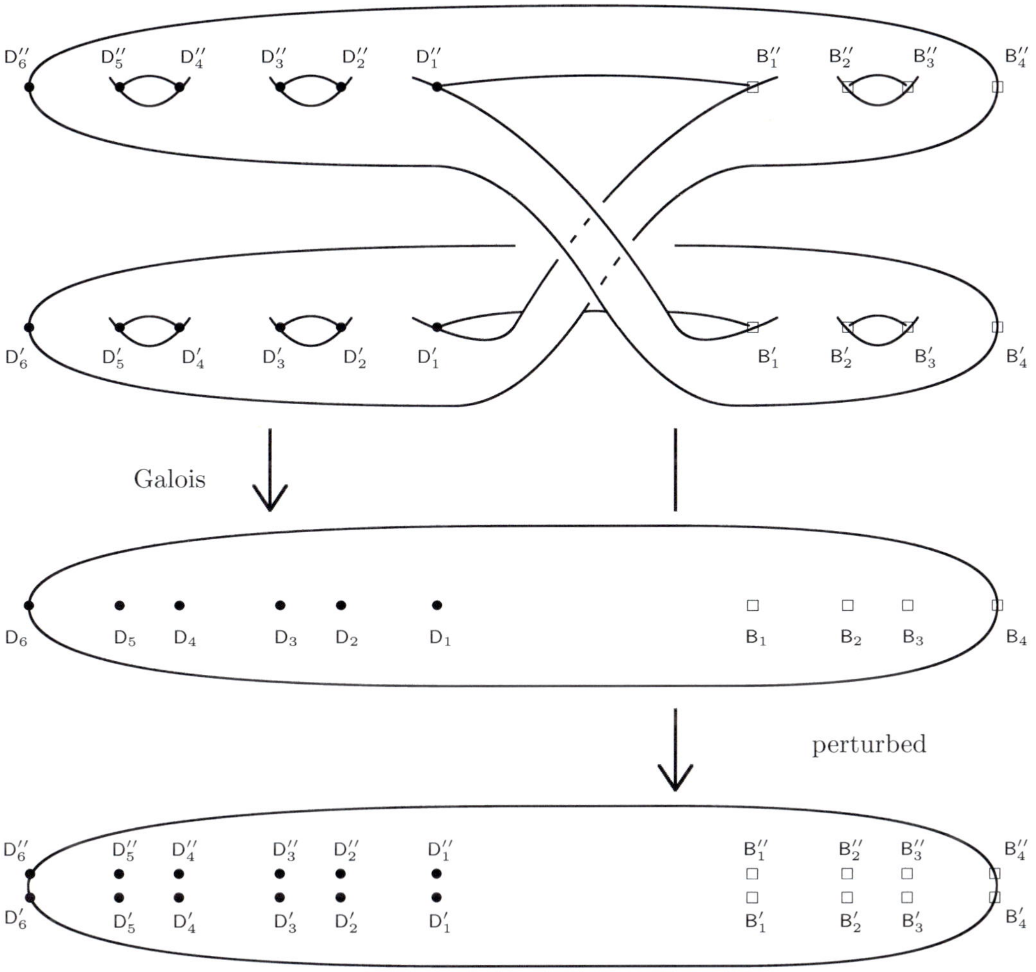
D″6
D″5
D″4
D″3
D″2
D″1
B″1
B″2
B″3
B″4
D′6
D′5
D′4
D′3
D′2
D′1
B′1
B′2
B′3
B′4
Galois
D6
D5
D4
D3
D2
D1
B1
B2
B3
B4
perturbed

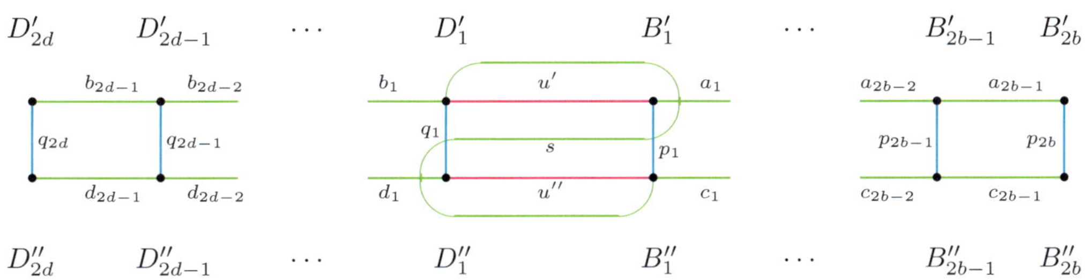
D′2d
D′2d−1
D′1
B′1
B′2b−1
B′2b
b2d−1
b2d−2
q2d
q2d−1
d2d−1
d2d−2
b1
u′
a1
q1
s
p1
d1
u″
c1
a2b−2
a2b−1
p2b−1
p2b
c2b−2
c2b−1
D″2d
D″2d−1
D″1
B″1
B″2b−1
B″2b

*The factorization is such that*

(1) *each* ($\pm$)*full-twist factor is of type* $p$ *or* $q$,
(2) *the weighted count of* ($\pm$) *full-twist factors of type* $p$ *yields* $8ab - 2(ad + bc)$,
(3) *the weighted count of* ($\pm$) *full-twist factors of type* $q$ *yields* $8cd - 2(ad + bc)$.

The result relies on a complete description of the braid monodromy factorization class associated to $\Delta$: this is given in Theorem 6.6, which proves indeed much more than what we need for the present purposes.

The most important application is the following Main Theorem:

**Theorem 6.4.** *Assume that* $S$ *is an abcd-surface,* $S'$ *is an* $a'b'c'd'$*-surface, that* $S$ *and* $S'$ *are homeomorphic, and that their vertical Braid monodromy factorizations are stably equivalent. Then, up to swapping the pair* $(a, b)$ *with the pair* $(c, d)$*, we have*

$$a + c = a' + c', b + d = b' + d', ab = a'b', cd = c'd'.$$

*The vertical Braid monodromy factorizations associated to an abc-surface* $S$ *and to an* $a'b'c'$*-surface* $S'$ *are not stably equivalent, except in the trivial cases* $a = a', b = b', c = c'$ *or* $a = c', b = b', c = a'$.

*Proof.* We sketch only the first calculation, which is not contained in [CLW].

Since $S$ and $S'$ are homeomorphic, they have the same $\chi$ and $K^2$, thus first of all $ab + cd = a'b' + c'd'$. By the previous theorem 6.3 and the forthcoming proposition (7.6) that thc factors of type $p$ are not stably conjugate to the factors of type $q$, it follows that $ab = a'b', cd = c'd'$.

Then by the same formulae of theorem 6.3 $(a + c)(b + d) = (a' + c')(b' + d')$, hence, since $K_S^2 = (a + c - 2)(b + d - 2)$, we get $(a + c) + (b + d) = (a' + c') + (b' + d')$, and therefore $a + c = a' + c', b + d = b' + d'$, up to swapping the pairs.

*Q.E.D.*

We start now to prepare the notation for the factorization theorem.

**Proposition 6.5.** *Up to homotopy in small discs containing pairs* $D_i', D_i''$ *or* $B_j', B_j''$ *the punctured fibre* $F_0$ *is given by the following picture*

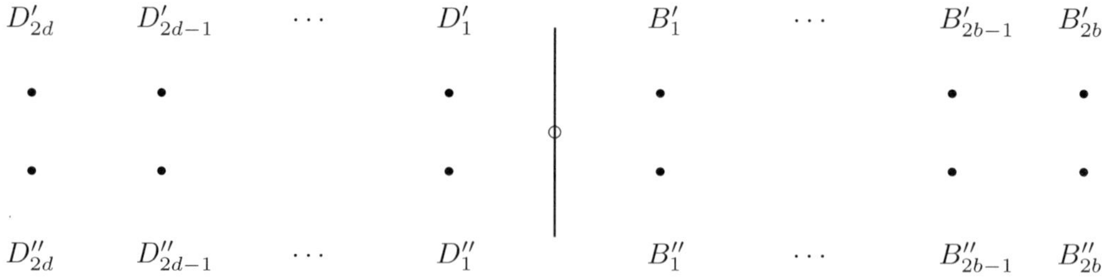


*and the covering monodromy* $\mu$ *of the perturbed bidouble cover, with respect to the origin, is given by*

$$\begin{array}{lllllll} \omega_{D'_j} & \mapsto & (12), & \qquad & \omega_{B'_j} & \mapsto & (13), \\ \omega_{D''_j} & \mapsto & (34), & \qquad & \omega_{B''_j} & \mapsto & (24), \end{array}$$

*where $\omega_P$ is any simple closed path around $P$ not crossing the imaginary axis $i\mathbb{R} \subset F_0$.*

*Proof.* As we remarked before, in the unperturbed Galois cover case we have real points $D_i$, $B_j$, only, the $D_i$'s with positive real coordinate, the $B_j$'s with negative real coordinates. And the covering's monodromy is (12)(34), resp. (13)(24) for paths which do not cross the imaginary axis.

The deformation then splits each branch point into two, hence the corresponding monodromies (which are transpositions) must be (12) and (34), resp. (13) and (24) (which commute). After a suitable homotopy the points are in the positions given by the picture. □

Referring to the above figures and to the figure below we introduce the following notation for a system of arcs, which are uniquely determined (up to homotopy) by their endpoints and the property that they are monotonous in the real coordinate and do not pass below any puncture; i.e., if they share the real coordinate with a puncture they have larger imaginary coordinate.

$$\begin{array}{llcll} p_i: & B'_i, B''_i, & \qquad & q_i: & D'_i, D''_i, \\ a_{ij}: & B'_i, B'_j, & & b_{ij}: & D'_i, D'_j, \\ c_{ij}: & B''_i, B''_j, & & d_{ij}: & D''_i, D''_j, \\ u'_{ij}: & B'_i, D'_j & & u''_{ij}: & B''_i, D''_j \end{array}$$

Further, there are arcs $s_{ij}$ connecting $D'_i$ with $B''_j$ according to a pattern which is shown in Figure 3.

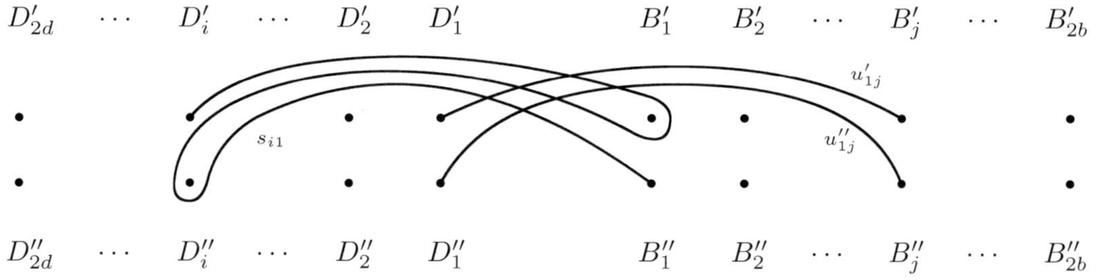


FIGURE 3. arcs for the theorem

**Theorem 6.6.** *The factorization is a product of factorizations*

$$\underbrace{\left(\beta_f \circ \underbrace{(\sigma_{p_1}^{\pm 2} \circ \cdots \circ \sigma_{p_1}^{\pm 2})}_{|2b-d|\ factors} \circ \beta_{fg}\right) \circ \cdots}_{2a\ repetitions} \circ \underbrace{(\sigma_{p_1}^{\pm 2} \circ \cdots \circ \sigma_{p_{2b}}^{\pm 2})}_{|2a-c|\ repetitions} \circ \cdots$$

$$\circ \underbrace{(\sigma_{q_1}^{\pm 2} \circ \cdots \circ \sigma_{q_{2d}}^{\pm 2}) \circ \cdots}_{|2c-a|\ repetitions} \circ \underbrace{\left(\beta_g \circ \underbrace{(\sigma_{q_1}^{\pm 2} \circ \cdots \circ \sigma_{q_1}^{\pm 2})}_{|2d-b|\ factors} \circ \beta_{gf}\right) \circ \cdots}_{2c\ repetitions}$$

*where the sign of the exponents 2 is constant inside a pair of brackets and is the sign of the number which determines the number of factors, ie.* $(2b-d)$ *resp.* $(2a-c)$, $(2c-a)$ *or* $(2d-b)$, *and where the* $\beta$*'s further decompose as products of factorizations*

$$\begin{aligned} \beta_f &= \beta_{f,2} \circ \cdots \circ \beta_{f,2b}, & \beta_{fg} &= \beta_{fg,2d} \circ \cdots \circ \beta_{fg,1}, \\ \beta_g &= \beta_{g,2} \circ \cdots \circ \beta_{g,2d}, & \beta_{gf} &= \beta_{gf,2b} \circ \cdots \circ \beta_{gf,1}, \end{aligned}$$

*based on elementary factorizations each having four factors*

$$\begin{aligned} \beta_{f,i} &= \sigma_{a_{1i}} \circ \sigma_{p_1}^2 \sigma_{c_{1i}} \sigma_{p_1}^{-2} \circ \sigma_{a_{1i}} \circ \sigma_{p_1}^2 \sigma_{c_{1i}} \sigma_{p_1}^{-2}, & \beta_{fg,j} &= \sigma_{u_{1j}}^3 \circ \sigma_{s_{1j}} \circ \sigma_{u'_{1j}}^3 \circ \sigma_{u''_{1j}}^3, \\ \beta_{g,j} &= \sigma_{b_{1j}} \circ \sigma_{q_1}^2 \sigma_{d_{1j}} \sigma_{q_1}^{-2} \circ \sigma_{b_{1j}} \circ \sigma_{q_1}^2 \sigma_{d_{1j}} \sigma_{q_1}^{-2}, & \beta_{gf,i} &= \sigma_{u_{i1}}^3 \circ \sigma_{s_{i1}} \circ \sigma_{u'_{i1}}^3 \circ \sigma_{u''_{i1}}^3. \end{aligned}$$

*The elementary factorizations originate in the regeneration of nodes of the branch curve of the corresponding bidouble Galois-cover.*

While it is complicated to explain in detail the above theorem, we would like to point out some other important tools in the local analysis of 'regenerations' of the singularities of the branch curve of the bidouble cover.

In the case of a 'proper node' ( i.e., a point where $f = g = 0$) we get a cusp-cluster, four critical points of which three are cusps and the last is a vertical tangency point.

The braid monodromy for the regeneration into a cusp-cluster had been thoroughly investigated before in [CW05], where the braid monodromy factorization type had been determined up to Hurwitz equivalence and simultaneous conjugation.

In [CLW] we were able to show how one can determine the Hurwitz equivalence class of the factorization from the datum of the product of the factors.

**Proposition 6.7.** *The braid monodromy factorization of a (regenerated) cusp-cluster with product* $\sigma_2^3\sigma_1\sigma_3\sigma_2\sigma_1^2\sigma_3^2$ *is Hurwitz-equivalent to the factorization*

$$\sigma_2^3 \circ \sigma_1\sigma_3\sigma_2\sigma_3^{-1}\sigma_1^{-1} \circ \sigma_1^3 \circ \sigma_3^3.$$

In the case of an improper node (i.e., a node of $f^\times = 0$ or of $g^\times = 0$) we get a cluster of vertical tangents, four critical points, all of which are vertical tangency points.

**Proposition 6.8.** *The braid monodromy factorization associated to a cluster of tangents corresponding to an improper node is given by*

$$\sigma_2\sigma_3\sigma_2^{-1} \circ \sigma_1\sigma_2\sigma_1^{-1} \circ \sigma_2\sigma_3\sigma_2^{-1} \circ \sigma_1\sigma_2\sigma_1^{-1}$$

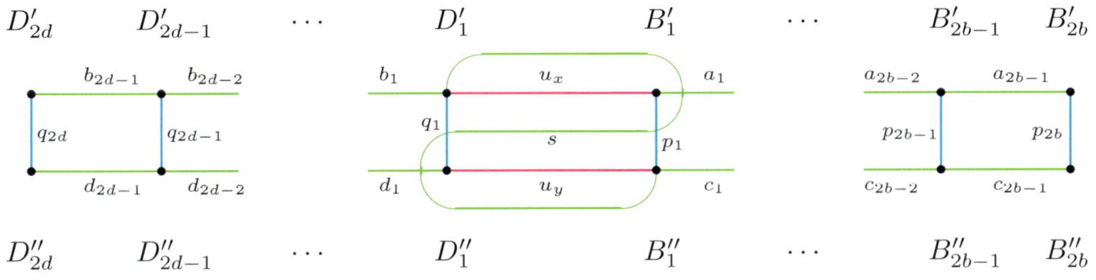


FIGURE 4. isotoped generators

*up to Hurwitz equivalence and simultaneous conjugation.*

## 7. Another proof of the non conjugacy theorem

This section is based on the classical correspondence between 4-tuple covers and triple covers, given by the surjection $\mathfrak{S}_4 \to \mathfrak{S}_3$ whose kernel is the Klein group $(\mathbb{Z}/2)^2$.

This idea was also used in [CLW] to obtain a triple cover $D_0 \to F_0$ corresponding to a quadruple cover $C_0 \to F_0$ and a resulting homomorphism of its mapping class group to the symplectic group acting on the $\mathbb{Z}/2$ homology.

In this way in the article [CLW] we gave a proof that the elements $\sigma_p^2$ and $\sigma_q^2$ in the braid monodromy group $H$ are not conjugate even in the stabilized braid monodromy group $\hat{H}$.

Observe that in the braid group there is a unique square root $\sigma$ of a full twist: since a full twist determines the homotopy class of an arc between two punctures, and this arc determines a half twist. In the braid group of the sphere, this square root is not unique (as pointed out by a referee), since another square root is also given by $\sigma\Delta^2$, where $\Delta^2$ is a generator of the centre, and has order 2. However, there is a unique square root in the conjugacy class of half-twists.

Hence it suffices to show that there is no $h$ in $\hat{H}$ such that $\sigma_p h = h\sigma_q$.

Our proof there exploited the action of the group $\hat{H}^+$ generated by $\hat{H}, \sigma_p$ and $\sigma_q$ on the $\mathbb{Z}/2\mathbb{Z}$ homology of a the triple cover $D_0$ of $F_0 \cong \mathbb{P}^1$.

In the alternative proof we are going to present, we look instead at the action of $\hat{H}^+$ on factorizations in $\mathfrak{S}_4$ of length $n = 4b + 4d$, restriction of the Hurwitz action of the braid group $\mathrm{Br}_n$.

In fact it suffices to consider the $\hat{H}^+$-orbit $\mathbb{O}$ of the factorization

$$\tau^0 : \ (12) \circ (34) \circ (12) \circ (34) \cdots (12) \circ (34) \circ (13) \circ (24) \circ (13) \circ (24) \cdots (13) \circ (24),$$

with $2b$ factors equal to $(12)$, respectively to $(34)$, and $2d$ factors equal to $(13)$, respectively to $(24)$. Note that $\tau^0$ is the factorization sequence corresponding to the covering monodromy for the 4 : 1 cover given by the restriction of our bidouble cover to the vertical projective line lying over the origin $x = 0$.

The orbit $\mathbb{O}$ is rather small by the following lemma.

**Lemma 7.1.** *The orbit* $\mathbb{O}$ *is contained in the set* $\hat{\mathbb{O}}$ *of factorizations made of transpositions*

$$\tau_1 \circ \tau_2 \circ \cdots \circ \tau_n$$

*such that for* $i \leq 4b$ *either* $\tau_i = (12)$ *or* $\tau_i = (34)$, *and for* $i > 4b$ *either* $\tau_i = (13)$ *or* $\tau_i = (24)$.

*Proof.* We must show that all elements in $\hat{H}^+$ leave invariant the set given in the claim.

To this purpose we exploit the surjection $\mathfrak{S}_4 \to \mathfrak{S}_3$. It maps transpositions to transpositions. Naturally the induced map on factorizations commutes with the Hurwitz action.

The upshot is now that the factorizations in the lemma are exactly the factorizations whose factors are transpositions and which map to one and the same factorization in $\mathfrak{S}_3$, namely

$$\bar{\tau}^0 : \ (12) \circ (12) \circ (12) \circ \cdots \circ (12) \circ (13) \circ (13) \circ (13) \circ \cdots \circ (13).$$

All generators of the braid monodromy subgroup $H$ act trivially on the monodromy factorization in $\mathfrak{S}_4$, hence also on the corresponding factorization $\bar{\tau}^0$ in $\mathfrak{S}_3$. Likewise do the elements $\sigma_p, \sigma_q$: since for instance $\sigma_q$ is the half twist connecting two neighbouring branch points whose respective local monodromies in $\mathfrak{S}_4$ yield consecutive factors $(1, 2)$ and $(3, 4)$: thus $\sigma_q$ exchanges these two factors.

It remains to prove that also the additional elements needed to generate $\hat{H}$ act trivially.

By definition of stable equivalence such additional generators are full-twists $\sigma_\alpha^2$ which preserve the monodromy factorization $\tau^0$ in $\mathfrak{S}_4$, hence also $\bar{\tau}^0$ and our claim follows. □

We record now in detail how our generators of $\hat{H}^+$ do act.

**Lemma 7.2.** *The following braids act trivially on all elements in* $\hat{\mathbb{O}}$:

$$\sigma_{u'}^3, \sigma_{u''}^3,$$

*and all additional generators of* $\hat{H}$ *which are not in* $H$.

*Proof.* Let us use as before the plane lexicographic ordering for the branch points.

Then one can easily verify that the element $\sigma_{4d}^3$ acts trivially on all factorizations, since $\tau_{4d}$ and $\tau_{4d+1}$ are always non commuting transpositions.

Hence both elements $\sigma_{u'}^3, \sigma_{u''}^3$, which are $H$-conjugate to $\sigma_{4d}^3$, act both trivially.

Consider now an additional generator $\sigma_\alpha^2$. It preserves $\tau^0$ and may be written as $\beta\sigma_1^2\beta^{-1}$.

Hence $\sigma_1^2$ acts trivially on $\bar{\tau}^0\beta$. Therefore the first two factors of $\bar{\tau}^0\beta$ must commute. Since both belong to $\mathfrak{S}_3$ they are even equal. The corresponding factors in any lift of $\bar{\tau}^0\beta$ to a factorization of transpositions in $\mathfrak{S}_4$ are then either equal or disjoint, so $\sigma_1^2$ acts trivially also on all such lifts.

We conclude that $\sigma_\alpha^2 = \beta\sigma_1^2\beta^{-1}$ acts trivially on all elements in $\hat{\mathbb{O}}$. □

**Lemma 7.3.** *The following types of braids act by the corresponding transpositions on factorizations in $\hat{\mathbb{O}}$:*

$$\sigma_a, \sigma_c, \sigma_b, \sigma_d, \sigma_p, \sigma_q.$$

*Proof.* We observe that all elements $\sigma_i$ with $i \neq 4d$ act by the corresponding transposition on the factorizations in $\hat{\mathbb{O}}$, because consecutive transpositions commute except for the transpositions in positions $4d$ and $4d+1$.

Hence the elements $\sigma_a, \sigma_c, \sigma_b, \sigma_d, \sigma_p$ and $\sigma_q$, which can be written as a product of factors different from $\sigma_{4d}$ act by the corresponding transpositions. □

**Lemma 7.4.** *The action of the 'snake' half twist $\sigma_s$ on factorizations in $\hat{\mathbb{O}}$ is given as follows:*

(1) *If the composition $\tau_{4d-1}\tau_{4d}\tau_{4d+1}\tau_{4d+2}$ is trivial or equal to $\pi = (14)(23)$, then $\sigma_s$ acts trivially.*

(2) *Otherwise $\sigma_s$ acts by* $\tau_i \mapsto \begin{cases} \pi\tau_i\pi & \textit{if } 4d-1 \leq i \leq 4d+2 \\ \tau_i & \textit{else} \end{cases}$

*Proof.* The 'snake' half twist $\sigma_s$ is given in terms of the Artin generators as

$$\sigma_s = \sigma_{4d}(\sigma_{4d-1}^2\sigma_{4d+1}^2)\sigma_{4d}(\sigma_{4d+1}^{-2}\sigma_{4d-1}^{-2})\sigma_{4d}^{-1}$$

Hence the element $\sigma_s$ acts on the subfactorization $\tau_{4d-1} \circ \tau_{4d} \circ \tau_{4d+1} \circ \tau_{4d+2}$ and leaves the remaining parts unchanged. It thus suffices to know how $\sigma_s$ acts on all 16 possible subfactorizations.

First we record the action on four specific factorizations. We decompose the action into five steps, each consisting of one or two elementary Hurwitz moves. In the first step we act by $\sigma_{4d}$, then by $(\sigma_{4d-1}^2\sigma_{4d+1}^2)$, by $\sigma_{4d}$, by $(\sigma_{4d+1}^{-2}\sigma_{4d-1}^{-2})$ and finally by $\sigma_{4d}^{-1}$.

$$\begin{array}{llll}
 & (12)\circ(12)\circ(13)\circ(13) & & (12)\circ(34)\circ(13)\circ(24) \\
\mapsto & (12)\circ(23)\circ(12)\circ(13) & \mapsto & (12)\circ(14)\circ(34)\circ(24) \\
\mapsto & (23)\circ(13)\circ(13)\circ(23) & \mapsto & (14)\circ(24)\circ(24)\circ(23) \\
\mapsto & (23)\circ(13)\circ(13)\circ(23) & \mapsto & (14)\circ(24)\circ(24)\circ(23) \\
\mapsto & (12)\circ(23)\circ(12)\circ(13) & \mapsto & (12)\circ(14)\circ(34)\circ(24) \\
\mapsto & (12)\circ(12)\circ(13)\circ(13) & \mapsto & (12)\circ(34)\circ(13)\circ(24)
\end{array}$$

We check that these first two factorizations belong to case (1) of the lemma and remain indeed unchanged.

$$\begin{array}{llll}
 & (12)\circ(34)\circ(13)\circ(13) & & (12)\circ(12)\circ(13)\circ(24) \\
\mapsto & (12)\circ(14)\circ(34)\circ(13) & \mapsto & (12)\circ(23)\circ(12)\circ(24) \\
\mapsto & (14)\circ(24)\circ(13)\circ(14) & \mapsto & (23)\circ(13)\circ(24)\circ(14) \\
\mapsto & (14)\circ(13)\circ(24)\circ(14) & \mapsto & (23)\circ(24)\circ(13)\circ(14) \\
\mapsto & (34)\circ(14)\circ(12)\circ(24) & \mapsto & (34)\circ(23)\circ(34)\circ(13) \\
\mapsto & (34)\circ(12)\circ(24)\circ(24) & \mapsto & (34)\circ(34)\circ(24)\circ(13)
\end{array}$$

The third and fourth factorizations belong to case (2) of the lemma. These subfactorizations are changed in all four positions. Every factor (12) is replaced by (34) and vice versa, and the same occurs with (13) and (24). But this amounts to the same result as conjugation of each factor by $\pi$.

The claim can be checked for the remaining 12 subfactorisations by elementary computations of the same kind, or by the observation that they correspond to the cases above under conjugation by $\pi$, by (12)(34), or by (13)(24). □

To define a suitable invariant of a factorization in $\hat{\mathbb{O}}$, we observe that each factor of a factorization $\tau$ in $\hat{\mathbb{O}}$ coincides either with the corresponding element of the factorization $\tau^0$ or with its conjugate by $\pi$.

The product of the factorization is left invariant by the braid group action, whence we claim that

$$\#\{i \mid \tau_i \neq \tau_i^0\}/2 \quad \in \quad \mathbb{N},$$

and thus we get a well-defined parity invariant of the $\hat{H}$-action.

The claim is easily established in this way: changing one of the factors in the right half amounts (in view of the commutativity of those factors) to multiplication by $(1,3)(2,4)$ on the right, while changing one of the factors in the left half amounts to multiplication by $(1,2)(3,4)$ on the left. Since $(1,2)(3,4)z(1,3)(2,4) \neq z \;\forall z \in \mathfrak{S}_4$ we conclude that the product is left unchanged only if an even number of changes are made.

**Lemma 7.5.** *The action of $\hat{H}$ on elements in $\hat{\mathbb{O}}$ preserves the parity of the integer*

$$M \quad := \quad \#\{i \mid \tau_i \neq \tau_i^0\}/2 \;+\; \#\{i > 4d,\ i \equiv 0 \pmod 2 \mid \tau_i \neq \tau_i^0\}$$

*Proof.* All generators of $\hat{H}$ except $\sigma_s$ act either trivially or by transpositions of factors on the same side (left or right) and at even distance, hence they preserve the first summand and likewise the second summand.

The action of $\sigma_s$ is different. We have the two possibilities of lemma 7.4. In the first case $\sigma_s$ acts trivially on $\tau$ and thus on $M$.

In the second case the product $\tau_{4d-1}\tau_{4d}\tau_{4d+1}\tau_{4d+2}$ is equal to (12)(34) or (13)(24) which implies, since in the Klein group ($\cong (\mathbb{Z}/2)^2$) a product of two non trivial elements is non trivial, that this part of the factorization is one of the following factorizations:

$$(12)\circ(34)\circ(13)\circ(13), \quad (12)\circ(34)\circ(24)\circ(24),$$
$$(34)\circ(12)\circ(13)\circ(13), \quad (34)\circ(12)\circ(24)\circ(24),$$

resp.

$$(12)\circ(12)\circ(13)\circ(24), \quad (12)\circ(12)\circ(24)\circ(13),$$
$$(34)\circ(34)\circ(13)\circ(24), \quad (34)\circ(34)\circ(24)\circ(13).$$

In all cases it can be checked that $\#\{i \in \{4d-1, 4d, 4d+1, 4d+2\} \mid \tau_i \neq \tau_i^0\}$ is either 1 or 3 and $\#\{i = 4d+2 \mid \tau_i \neq \tau_i^0\}$ is either 0 or 1.

Both numbers change when the factorization is conjugated by $\pi$, hence the parity of $M$ is preserved, as claimed.

□

Finally we can prove the essential step:

**Proposition 7.6.** *The elements $\sigma_p^2$ and $\sigma_q^2$ are not conjugate in $\hat{H}$. Hence also $\sigma_p$ and $\sigma_q$ are not conjugate in $\hat{H}$.*

*Proof.* Suppose on the contrary that there exists an $h \in \hat{H}$ such that $\sigma_p h = h\sigma_q$. Then the action of both sides on the factorization $\tau^0$ must give the same factorization. But the invariant $M$ is 2 for $\tau^0 h\sigma_q = \tau^0\sigma_q$ and odd for $\tau^0\sigma_p h$, since it is 1 for $\tau_0\sigma_p$. So the factorizations are different and we get a contradiction.

The second assertion was already shown at the beginning of the present section.

□

## 8. Non equivalence of the horizontal Lefschetz fibrations in some cases.

Consider an *abc*-surface $S$ and its vertical Lefschetz fibration $F : S \to \mathbb{P}^1$ given by the composition of the perturbed covering $\pi : S \to Q$ with the first projection onto $\mathbb{P}^1$; the differentiable class of $F$ is encoded in the M-equivalence class of the factorization of the identity in the Mapping class group $\mathcal{M}ap_g$ corresponding to the monodromy factorization of $F$ ($g$ is the genus of the base fibre $C_0$ of $F$) .

As already mentioned, we proved in [CW04] that if $S'$ is an $(a', b', c')$ surface, where $b = b'$ and $a + c = a' + c'$ then $S$ and $S'$ are not deformation equivalent but the above defined Lefschetz fibrations are M-equivalent, in particular proving that the surfaces $S$ and $S'$ are diffeomorphic via an orientation preserving diffeomorphism sending the canonical class to the canonical class.

We want to give here another small contribution in the direction of the interesting and difficult question whether the surfaces are symplectomorphic with respect to the canonical symplectic structure induced by the complex structure.

The projections of $S$ and $S'$ to $Q = \mathbb{P}^1 \times \mathbb{P}^1$ have discriminant curves $\Delta$ and $\Delta'$. Our computations ([CLW]) of the Braid monodromy factorizations of these curves with respect to the vertical projection (onto the $x$-axis), that we illustrated previously, show that the braid monodromy factorizations are not stably equivalent. This rises the suspicion that the surfaces may be not canonically symplectomorphic.

We shall in fact point out another difference between the surfaces $S$ and $S'$. We consider the Lefschetz fibrations corresponding to the horizontal projection (to the $y-axis$). This time a regular fibre $C_0'$ is a 4-tuple covering of a horizontal line in $Q$, with (4a+4c) branch points, hence it has genus 2a+2c-3. It is identified in a natural way with the curve on Figure 5.

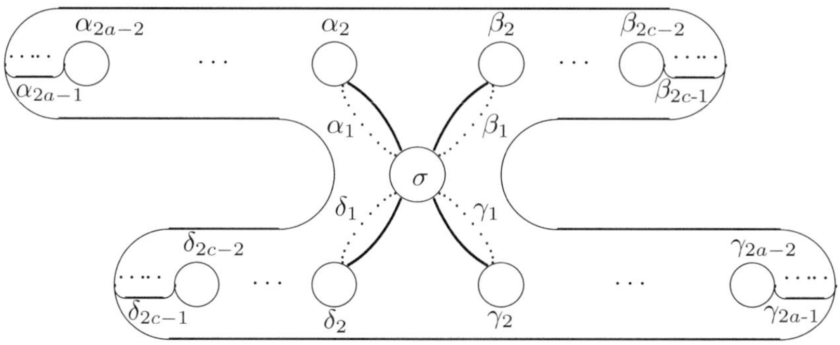


FIGURE 5. The fibre $C'_0$ of the horizontal Lefschetz fibration

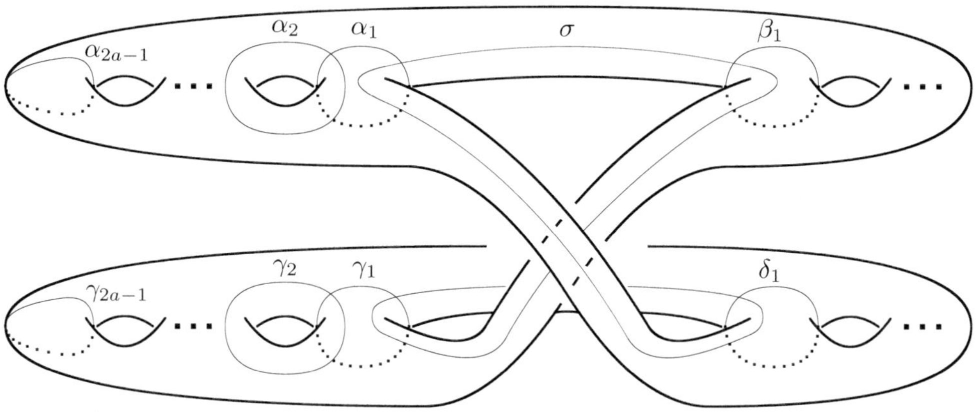


FIGURE 6

Our computation of the Braid monodromy of the discriminant curve, where the roles of the variables $x$ are easily exchanged, yields easily the monodromy factorization corresponding to the horizontal Lefschetz fibration. In particular it shows that the monodromy group of the Lefschetz fibration is generated by Dehn twists with respect to the cycles $\alpha_i, \beta_i, \gamma_i, \delta_i, \sigma$ on Figure 6. We shall prove that if $a + c = a' + c'$ is even and $a - a'$ is odd and $b = b'$ then the monodromy factorizations for the horizontal Lefschetz fibrations are not M-equivalent although the fibrations have the same genus and the surfaces $S$ and $S'$ are diffeomorphic.

We map the cycles in $C'_0$ into the homology group $H_1(C'_0, \mathbb{Z})$ and further into the vector space $V = H_1(C'_0, \mathbb{Z}/2\mathbb{Z})$ of vectors modulo 2. The intersection form of cycles on

$C'_0$ induces an alternating symmetric form $(u, v)$ on $V$. The mapping class group of $C'_0$ projects onto the symplectic group of $V$ with respect to the form and Dehn twists project to transvections $T$ with respect to cycles: $T_u(v) = v + (u, v)u$.

We shall denote the images of $\alpha_i, \beta_i, \gamma_i, \delta_i, \sigma$ in $V$ by $a_i, b_i, c_i, d_i, s$ respectively.

Considering the intersection matrix of these vectors we see that if we omit the vectors $b_{2c-1}, c_{2a-1}, d_{2c-1}$ then the remaining vectors have a non-singular intersection matrix and thus they form a basis of $V$.

We define a quadratic function $q : V \to \mathbb{Z}/2\mathbb{Z}$ using this basis:

$q$ is equal 1 on each basis vector and $q(u + v) = q(u) + q(v) + (u, v)$ for any pair of vectors $u, v$.

Quadratic functions with an associated bilinear form as above fall into two equivalence classes according to their Arf invariant (modulo 2). In order to compute the Arf invariant we choose a symplectic basis $e_i, f_i$ of $V$, such that $(e_i, f_i) = 1$ for all $i$, while all other intersections products of basis vectors are equal 0. Then the Arf invariant of $q$ is equal to $\Sigma_i q(e_i)q(f_i)$.

We now recall Theorem 4 from [Waj80]:

**Theorem 8.1.** *Let $G$ be a subgroup of $Sp(V, \mathbb{Z}/2\mathbb{Z})$ generated by transvections, containing the transvections with respect to the vectors $u_i$ of a fixed basis. Let $q$ be the quadratic function defined by $q(u_i) = 1$.*

*If $G$ contains a transvection with respect to a vector $v$ satisfying $q(v) = 0$ then $G = Sp(V, \mathbb{Z}/2\mathbb{Z})$.*

*If this is not the case and the basis is not special, e.g. the intersection diagram of the basis is a tree but it is different from the Dynkin diagrams $A_n$ and $D_n$, then $G$ is equal to the orthogonal group of the function $q$, the group of all elements of $Sp(V, \mathbb{Z}/2\mathbb{Z})$ preserving $q$.*

Let us go back to the Lefschetz fibration of the surface $S$. Let $G$ be the image of the monodromy group, the group generated by the transvections with respect to the vectors $a_i, b_i, c_i, d_i, s$. Consider the basis described before. The intersection diagram of this basis has the form of a cross, thus the basis is not special.

Consider the corresponding quadratic function $q$. The vector $b_{2c-1}$, which is not in the basis, is equal to the sum $v = a_1 + a_3 + a_5 + \cdots + a_{2a-1} + b_1 + b_3 + b_5 + \cdots + b_{2c-3}$ (both vectors have the same intersection with each basis vector so they are equal).

If $a + c$ is odd then $q(v) = 0$ and the group $G$ generated by the transvections coincides with $Sp(V, \mathbb{Z}/2\mathbb{Z})$.

If $a + c$ is even then $q(b_{2c-1}) = q(c_{2a-1}) = q(d_{2c-1}) = 1$ and the group $G$ is equal to the orthogonal group of $q$. In this case if we start with any basis $v_i$ satisfying $q(v_i) = 1$ and we define a quadratic function using the basis $v_i$ we get the same function $q$.

It is easy to compute the Arf invariant of $q$.

To simplify notation we work out an example with $a = 2, c = 2$.

We can choose the following symplectic basis : $(a_3, a_2), (a_3 + a_1, s), (a_3 + a_1 + b_1, b_2)$, $(a_3 + a_1 + c_1, c_2), (a_3 + a_1 + d_1, d_2)$.

The sum of $k$ pairwise non-intersecting basis vectors has the $q$-value 1 if $k$ is odd and has the $q$-value 0 if $k$ is even.

Hence the Arf invariant of $q$ is 0 if $a$ is even and is 1 if $a$ is odd.

We now prove that if $a + c = a' + c'$ is even and $a - a'$ is odd then the monodromy factorizations of the horizontal Lefschetz fibrations of an abc-surface $S$ and of an a'b'c'-surface $S'$ are not M-equivalent. Without loss of generality assume that a is even, and that a' is odd: so that the Arf invariant of $q$ is 0, while the Arf invariant of $q'$ is 1.

Assume the contrary: then, in fact, making an arbitrary identification of the fibers of the fibrations, we may assume that we have two factorizations by Dehn twists which are Hurwitz equivalent.

We map the corresponding cycles to vectors in $V$. The monodromy factorization maps to a factorization in $Sp(V, \mathbb{Z}/2\mathbb{Z})$ by transvections with respect to some vectors.

This set of vectors contains a basis with the intersection pattern which defines a quadratic function $q$ with the Arf invariant 0. All vectors of the initial set have the $q$-value 1. The action induced by a Hurwitz move in $Sp(V, \mathbb{Z}/2\mathbb{Z})$ replaces a pair of transvections $T_u, T_v$ by the pair $T_v, T_v^{-1}T_uT_v$. One checks that $T_v^{-1}T_uT_v = T_w$ where $w = u + (u, v)v$. Assume $q(u) = q(v) = 1$: then $q(w) = 1$. So, after the Hurwitz move, we have again a product of transvections with respect to vectors of $q$-value 1. If we assume that the factorizations are M-equivalent, then we get a factorization by transvections with respect to a set of vectors having $q$-value 1 and which contains a basis with an intersection pattern such that it defines a quadratic function $q'$ with Arf invariant 1. But, by the previous remark , $q' = q$. This is a contradiction.

One can easily show that the orthogonal groups of $q$ and $q'$ are not conjugate in $Sp(V, \mathbb{Z}/2\mathbb{Z})$ and therefore the monodromy groups of the two fibrations are not conjugate in the mapping class group of a fibre. Probably these monodromy groups are not isomorphic at all, but we do not know at the moment how to prove it.

## 9. Epilogue

The question of determining whether the Braid Monodromy factorizations of two diffeomorphic surfaces $S$, $S'$ (an abc-surface, resp. an a'b'c'-surface) are M-equivalent was motivated by the related question of deciding about the existence of a diffeomorphism $\varphi$ between $S, S'$ commuting with the respective vertical Lefschetz fibrations, and yielding a canonical symplectomorphism.

Our main result that these factorizations are not stably equivalent suggests that such a diffeomorphism cannot exist.

The relation of stable equivalence (the authors call it indeed m-equivalence) was introduced by Auroux and Katzarkov in [A-K00] in order to obtain invariants of symplectic 4-manifolds.

The reason to introduce it, allowing not only Hurwitz equivalence and simultaneous conjugation, but also creation/cancellation of admissible pairs of a positive and of a negative node (here a node, and its corresponding full twist $\beta$, is said to be admissible if the inverse image of the node inside the ramification divisor consists of two disjoint smooth branches) has to do with the fact that this phenomenon occurs when considering isotopies of approximately holomorphic maps.

Our result (6.3) shows that the braid monodromy group $H$ depends only upon the numbers $b$ and $d$, provided for instance that we have nonvanishing of the respective numbers $8ab - 2(ad + bc)$, $8cd - 2(ad + bc)$, or provided that we are in the case $b = d$.

A fortiori if the groups $H$ are the same for different choices of $(a, b, c, d)$ then the fundamental groups $\pi_1(Q \setminus \Delta)$ are isomorphic.

Our new method of distinguishing factorizations, and not only associated fundamental groups of complements of branch curves (by the Zariski-van Kampen method), leads to the above results representing the first positive step towards the realization of a more general program set up by Moishezon ([Moi81], [Moi83]) in order to produce braid monodromy invariants which should distinguish the connected components of a moduli space $\mathfrak{M}_{\chi,K^2}$.

Moishezon's program is based on the consideration (assume here for simplicity that $K_S$ is ample) of a general projection $\Psi_m : S \to \mathbb{P}^2$ of a pluricanonical embedding $\Phi_m : S \to \mathbb{P}^{P_m - 1}$, and of the braid monodromy factorization corresponding to the (cuspidal) branch curve $B_m$ of $\Psi_m$.

An invariant of the connected component is here given by the M-equivalence class (i.e., for Hurwitz equivalence plus simultaneous conjugation) of this braid monodromy factorization. Moishezon, and later Moishezon-Teicher, calculated a coarser invariant, namely the fundamental group $\pi_1(\mathbb{P}^2 - B_m)$.

This group turned out to be not overly complicated, and in fact, as shown in many cases in [ADKY04], it tends to give no extra information beyond the one given by the topological invariants of $S$ (such as $\chi, K^2$).

Auroux and Katzarkov showed that, for $m >> 0$, the stable-equivalence class of the above braid monodromy factorization determines the canonical symplectomorphism class of $S$, and conversely.

The work by Auroux, Katzarkov adapted Donaldson's techniques for proving the existence of symplectic Lefschetz fibrations ([Don96-2],[Don99]) in order to show that each integral symplectic 4-manifold is in a natural way 'asymptotically' realized by a generic symplectic covering of $\mathbb{P}^2_{\mathbb{C}}$, given by approximately holomorphic sections of a high multiple $L^{\otimes m}$ of a complex line bundle $L$ whose class is the one of the given integral symplectic form.

The methods of Donaldson on one side, Auroux and Katzarkov on the other, use algebro geometric methods in order to produce invariants of symplectic 4-manifolds.

For instance, in the case of a generic symplectic covering of the plane, we get a corresponding branch curve $\Delta_m$ which is a symplectic submanifold with singularities only nodes and cusps.

To $\Delta_m$ corresponds then a factorization in the braid group, called m-th braid monodromy factorization: it contains, as in the present paper, only factors which are conjugates of $\sigma_1^j$, not only with $j = 1, 2, 3$ as in the complex algebraic case, but also with $j = -2$.

The factorization is not unique because it may happen that a pair of two consecutive nodes, one positive and one negative, may be created, or disappear, and this is the reason to consider its stable equivalence class which the authors show, for $m >> 1$, to be an invariant of the integral symplectic manifold.

As we have already remarked, in the case of abc-surfaces the vertical Braid monodromy groups $H$ are determined by $b$ (up to conjugation), hence the fundamental groups $\pi_1(Q \setminus \Delta)$ are isomorphic for a fixed value of $b$.

So, Moishezon's technique produces no invariants.

Our result indicates that one should try to go all the way in the direction of understanding stable-equivalence classes of pluricanonical braid monodromy factorizations. Let us describe here how this could be done.

Define $p : S \to \mathbb{P}^2$ as the morphism $p$ obtained as the composition of $\pi : S \to Q$ with the embedding $Q \hookrightarrow \mathbb{P}^3$ followed by a general projection $\mathbb{P}^3 \dashrightarrow \mathbb{P}^2$.

In the even more special case of abc-surfaces such that $a+c = 2b$, the m-th pluricanonical mapping $\Phi_m : S \to \mathbb{P}^{P_m - 1}$ has a (non generic) projection given by the composition of $p$ with a Fermat type map $\nu_r : \mathbb{P}^2 \to \mathbb{P}^2$ (given by $\nu_r(x_0, x_1, x_2) = (x_0^r, x_1^r, x_2^r)$ in a suitable linear coordinate system), where $r := m(2b - 2)$.

Let $B$ be the branch curve of a generic perturbation of $p$: then the braid monodromy factorization corresponding to $B$ can be calculated from the braid monodromy factorization corresponding to $\Delta$.

We hope, in a sequel to this paper, to be able to determine whether these braid monodromy factorizations are equivalent, respectively stably-equivalent, for abc-surfaces such that $a + c = 2b$.

The problem of calculating the braid monodromy factorization corresponding to the (cuspidal) branch curve $B_m$ starting from the braid monodromy factorization of $B$ has been addressed, in the special case $m = 2$, by Auroux and Katzarkov ([A-K08]). Iteration of their formulae should lead to the calculation of the braid monodromy factorization corresponding to the (cuspidal) branch curve $B_m$ in the case, sufficient for applications, where $m$ is a sufficiently large power of 2.

Whether it will be possible to decide about the stable equivalence of the factorizations which these formidable calculations will yield, is a quite open question.

In both directions the result would be extremely interesting, leading either to

i) a counterexample to the speculation DEF = CAN. SYMPL also in the simply connected case, or to

ii) examples of diffeomorphic but not canonically symplectomorphic simply connected algebraic surfaces.

**Acknowledgements:** The present research was performed in the realm of the Forschergruppe 790 'Classification of algebraic surfaces and compact complex manifolds' of the DFG (DeutscheForschungsGemeinschaft).

We are grateful to a referee for a very careful reading of the manuscript which saved us from many misprints and inaccuracies.

## References


[ABKP00] J. Amoros, F. Bogomolov, L. Katzarkov, T. Pantev, *Symplectic Lefschetz fibrations with arbitrary fundamental group*, With an appendix by Ivan Smith, J. Differential Geom. 54, no. 3, (2000), 489–545.

[Art26] E. Artin, *Theorie der Zöpfe,* Hamburg Univ. Math. Seminar Abhandlungen 4-5 (1926), 47–72.

[A-K00] D. Auroux, L. Katzarkov, *Branched coverings of* $\mathbb{CP}^2$ *and invariants of symplectic 4-manifolds* Inv. Math. 142 (2000), 631–673.

[Aur02] D. Auroux, *Fiber sums of genus 2 Lefschetz fibrations.* Turkish J. Math. 27, no. 1 (2003), 1–10.

[ADK03] D. Auroux, S. Donaldson, L. Katzarkov, *Luttinger surgery along Lagrangian tori and non-isotopy for singular symplectic plane curves.* Math. Ann. 326 no. 1 (2003), 185–203.

[ADKY04] D. Auroux, S. Donaldson, L. Katzarkov and M. Yotov, *Fundamental groups of complements of plane curves and symplectic invariants* Topology 43, no. 6 (2004), 1285–1318.

[A-K08] D. Auroux, L. Katzarkov , *A degree doubling formula for braid monodromies and Lefschetz pencils*, Pure Appl. Math. Q. 4, no. 2, part 1, (2008) 237–318.

[BCG07] I. Bauer, F. Catanese, F. Grunewald, *The absolute Galois group acts faithfully on the connected components of the moduli spaces of surfaces of general type,* arXiv:0706.1466.

[BCG08] I. Bauer, F. Catanese, F. Grunewald, *The classification of surfaces with* $p_g = q = 0$ *isogenous to a product of curves,* math.AG/0610267, Pure Appl. Math. Q. 4, no. 2, part 1, 547–586 (2008).

[Bir74] J.S. Birman, *Braids, links, and mapping class groups,* Annals of Mathematics Studies, No. 82, Princeton University Press, Princeton, N.J.; University of Tokyo Press, Tokyo (1974).

[Bom73] E. Bombieri, *Canonical models of surfaces of general type,* Publ. Math. I.H.E.S., 42 (1973), 173–219.

[Cat82] F. Catanese, *Moduli of surfaces of general type.* in 'Algebraic geometry–open problems' (Ravello, 1982), Lecture Notes in Math., 997 Springer, Berlin-New York (1983), 90–112.

[Cat84] F. Catanese, *On the Moduli Spaces of Surfaces of General Type,* J. Differential Geom. 19 (1984), 483–515.

[Cat87] F. Catanese, *Automorphisms of Rational Double Points and Moduli Spaces of Surfaces of General Type,* Comp. Math. 61 (1987), 81–102.

[Cat86] F. Catanese, *Connected Components of Moduli Spaces,* J. Differential Geom. 24 (1986), 395–399.

[Cat88] F. Catanese, *Moduli of algebraic surfaces.* Theory of moduli (Montecatini Terme, 1985), Lecture Notes in Math. 1337, Springer, Berlin (1988), 1–83.

[Cat00] F. Catanese, *Fibred surfaces, varieties isogenous to a product and related moduli spaces,* Amer. J. Math. 122, no.1 (2000), 1–44.

[Cat03] F. Catanese, *Moduli Spaces of Surfaces and Real Structures,* Ann. Math.(2) 158, no.2 (2003),577–592.

[Cat02] F. Catanese, *Symplectic structures of algebraic surfaces and deformation,* 14 pages, math.AG/0207254.

[CW04] F. Catanese, B. Wajnryb, *Diffeomorphism of simply connected algebraic surfaces.* math.AG/0405299 , J. Differ. Geom. 76, No. 2, 177–213 (2007).

[CW05] F. Catanese, B. Wajnryb, *The 3-cuspidal quartic and braid monodromy of degree 4 coverings.* Ciliberto, C. (ed.) et al., Projective varieties with unexpected properties. A volume in memory of Giuseppe Veronese. Proceedings of the international conference "Varieties with unexpected properties", Siena, Italy, June 8–13, 2004. Berlin: Walter de Gruyter. 113–129 (2005).

[Cat06] F. Catanese, *Canonical symplectic structures and deformations of algebraic surfaces,* math.AG/0608110, Comm. in Contemp. Math. 11 (2009), n. 3, 481–493.

[Cat08] F. Catanese, *Differentiable and deformation type of algebraic surfaces, real and symplectic structures.* (English) Catanese, Fabrizio (ed.) et al., Symplectic 4-manifolds and algebraic surfaces. Lectures given at the C.I.M.E. summer school, Cetraro, Italy, September 2–10, 2003. Berlin: Springer; Florence: Fondazione C.I.M.E. Lecture Notes in Mathematics 1938, 55–167 (2008).

[Cat09] F. Catanese, *Canonical symplectic structures and deformations of algebraic surfaces.* Commun. Contemp. Math. 11, No. 3, (2009) 481–493.

[CLW] F. Catanese, M. Lönne, B. Wajnryb, *Moduli spaces and braid monodromy types of bidouble covers of the quadric,* arXiv:0910.2142.

[Chi44] O. Chisini, *Sulla identità birazionale delle funzioni algebriche di due variabili dotate di una medesima curva di diramazione,* Ist. Lombardo Sci. Lett. Cl. Sci. Mat. Nat. Rend. (3) 8(77) (1944), 339–356.

[Chi55] O. Chisini, *Il teorema di esistenza delle trecce algebriche. I-III,* Atti Accad. Naz. Lincei. Rend. Cl. Sci. Fis. Mat. Nat. (8), 17 - 18(1954), 143–149; (1955), 307–311; (1955), 8–13.

[Dehn38] M. Dehn, *Die Gruppe der Abbildungsklassen. (Das arithmetische Feld auf Flächen.)* Acta Math. 69 (1938), 135–206.

[Don83] S.K. Donaldson, *An Application of Gauge Theory to Four-Dimensional Topology,* J. Differential Geom. 18 (1983), 279–315.

[Don86] S.K. Donaldson, *Connections, cohomology and the intersection forms of 4-manifolds.* J. Differential Geom. 24 (1986), 275–341.

[Don90] S.K. Donaldson, *Polynomial invariants for smooth four-manifolds.* Topology 29, 3 (1990), 257–315.

[Don92] S.K. Donaldson, *Gauge theory and four-manifold topology.* [CA] Joseph, A. (ed.) et al., First European congress of mathematics (ECM), Paris, France, July 6-10, 1992. Volume I: Invited lectures (Part 1). Basel: Birkhäuser, Prog. Math. 119 (1994) 121–151 .

[Don96] S.K. Donaldson, *The Seiberg-Witten Equations and 4-manifold topology.* Bull. Am. Math. Soc., (N S) 33, 1 (1996) 45–70.

[Don96-2] S.K. Donaldson, *Symplectic submanifolds and almost-complex geometry.* J. Differential Geom. 44, 4 (1996), 666–705.

[Don99] S.K. Donaldson, *Lefschetz pencils on symplectic manifolds,* J. Differential Geom. 53, no. 2 (1999), 205–236.

[D-K90] S.K. Donaldson, P.B. Kronheimer *The geometry of four-manifolds.* Oxford Mathematical Monographs. Oxford: Clarendon Press, (1990) ix, 440 pp.

[Ehr43] C. Ehresmann, *Sur les espaces fibrés différentiables.* C.R. Acad. Sci. Paris 224 (1947), 1611–1612.

[Enr49] F. Enriques, *Le Superficie Algebriche.* Zanichelli, Bologna (1949).

[Free82] M. Freedman, *The topology of four-dimensional manifolds.* J. Differential Geom. 17, n. 3 (1982), 357–454.

[F-Q90] M. Freedman and F.Quinn, *Topology of 4- manifolds.* Princeton Math. Series 39 (1990).

[F-M-M87] R. Friedman, B.G. Moishezon, J.W.Morgan, *On the $C^\infty$ invariance of the canonical classes of certain algebraic surfaces.* Bull. Amer. Math. Soc. 17 (1987), 283–286.

[F-M88] R. Friedman and J.W.Morgan, *Algebraic surfaces and four-manifolds: some conjectures and speculations.* Bull. Amer. Math. Soc. 18 (1988) 1–19.

[F-M94] R. Friedman; J. W. Morgan, *Smooth four-manifolds and complex surfaces,* Ergebnisse der Mathematik und ihrer Grenzgebiete (3), 27 Springer-Verlag, Berlin, (1994), x+520 pp.

[F-M97] R. Friedman and J.W.Morgan, *Algebraic surfaces and Seiberg-Witten invariants.* J. Algebr. Geom. 6, No.3 (1997) 445–479.

[Gie77] D. Gieseker, *Global moduli for surfaces of general type,* Invent. Math. 43, no. 3(1977), 233–282.

[Gom95] R.E. Gompf, *A new construction of symplectic manifolds* Ann. of Math. 142 3 (1995), 527–595.

[Gom01] R.E. Gompf, *The topology of symplectic manifolds.* Turkish J. Math. 25 no. 1, (2001), 43–59.

[G-S99] R.E. Gompf, A. Stipsicz, *4-manifolds and Kirby calculus.* Graduate Studies in Mathematics, 20. American Mathematical Society, Providence, RI, (1999) xvi+558 pp.
[GMW04] J. Gonzales-Meneses, B. Wiest, On the structure of the centralizer of a braid, Ann. Sci. École Norm. Sup. 37 no. 5 (2004), 729–757
[HT80] A. Hatcher, W.Thurston, *A presentation for the mapping class group of a closed orientable surface,* Topology 19, no. 3(1980), 221–237.
[Kas80] A. Kas, *On the handlebody decomposition associated to a Lefschetz fibration.* Pacific J. Math. 89 no. 1, (1980), 89–104.
[Kat83] *Open problems: Classification of algebraic and analytic manifolds.* Classification of algebraic and analytic manifolds, Proc. Symp. Katata/Jap. 1982. Edited by Kenji Ueno. Progress in Mathematics, 39. Birkhäuser, Boston, Mass. (1983), 591–630.
[KK02] V. M. Kharlamov, V.S. Kulikov, *On real structures of real surfaces.* Izv. Ross. Akad. Nauk Ser. Mat. 66, no. 1, 133–152 (2002); translation in Izv. Math. 66, no. 1, 133–150 (2002)
[KK02-b] V. Kharlamov, V. Kulikov, *Deformation inequivalent complex conjugated complex structures and applications,* Turkish J. Math. 26 , no. 1 (2002), 1–25.
[K-S58] K. Kodaira, D.C. Spencer, *On deformations of complex analytic structures. I, II,* Ann. of Math. (2) 67 (1958), 328–466.
[Kod66] K. Kodaira, *On the structure of compact complex analytic surfaces. II.* Amer. J. Math. 88 (1966), 682–721.
[K-SB88] J. Kollár and N.I. Shepherd Barron, *Threefolds and deformations of surface singularities.* Inv. Math. 91 (1988) 299–338.
[Kul99] V.S. Kulikov, *On Chisini's conjecture,* Izv. Ross. Akad. Nauk Ser. Mat. 63, no. 6(1999), 83–116 ; translation in Izv. Math. 63, no. 6 (1999), 1139–1170.
[Kul06] V.S. Kulikov, *On Chisini's conjecture. II* math. AG/0610356v1.
[M-M76] R. Mandelbaum, B. Moishezon, *On the topological structure of non-singular algebraic surfaces in $CP^3$,* Topology 15(1976), 23–40.
[M-M80] R. Mandelbaum, B. Moishezon, *On the topology of simply-connected algebraic surfaces,* Trans. Am. Math. Soc. 260 (1980), 195–222.
[Man94] M. Manetti, *On some Components of the Moduli Space of Surfaces of General Type,* Comp. Math. 92 (1994) 285–297.
[Man96] M. Manetti, *Degenerations of Algebraic Surfaces and Applications to Moduli Problems,* Tesi di Perfezionamento Scuola Normale Pisa (1996) 1–142.
[Man97] M. Manetti, *Iterated Double Covers and Connected Components of Moduli Spaces,* Topology 36, 3 (1997) 745–764.
[Man01] M. Manetti, *On the Moduli Space of diffeomorphic algebraic surfaces,* Inv. Math. 143 (2001), 29–76.
[Moi77] B. Moishezon, *Complex surfaces and connected sums of complex projective planes,* Lecture Notes in Math., 603, Springer, Berlin (1977)
[Moi81] B. Moishezon, *Stable branch curves and braid monodromies,* Algebraic geometry, Proc. Conf., Chicago Circle 1980, Lect. Notes Math. 862, 107–192 (1981).
[Moi83] B. Moishezon, *Algebraic surfaces and the arithmetic of braids I,* in 'Arithmetic and geometry, Pap. dedic. I. R. Shafarevich, Vol. II: Geometry' Prog. Math. 36, Birkhäuser(1983), 199–269.
[MT92] B. Moishezon, M. Teicher, *Finite fundamental groups, free over $\mathbb{Z}/c\mathbb{Z}$, for Galois covers of* $\mathbb{CP}^2$. Math. Ann. 293, no. 4(1992), 749–766.
[Moi94] B. Moishezon, *The arithmetics of braids and a statement of Chisini,* in 'Geometric Topology, Haifa 1992', Contemp. Math. 164, Amer. Math. Soc., Providence, RI (1994), 151–175.
[Mor96] J. W. Morgan, *The Seiberg-Witten equations and applications to the topology of smooth four-manifolds.* Mathematical Notes 44. Princeton Univ. Press vi (1996).
[Mos65] J. Moser, *On the volume elements on a manifold.* Trans. Amer. Math. Soc. 120 (1965), 286–294.

[Ser64] J.P. Serre, *Exemples de variétés projectives conjuguées non homéomorphes.* C. R. Acad. Sci. Paris 258, (1964) 4194–4196.

[Thur76] W.P. Thurston, *Some simple examples of symplectic manifolds,* Proc. Amer. Math. Soc. 55, no. 2 (1976), 467–468.

[Waj80] B. Wajnryb, *On the Monodromy of Plane Curve Singularities,* Math. Ann. 246, 141–154 (1980).

[Waj83] B. Wajnryb, *A simple presentation for the mapping class group of an orientable surface,* Israel J. Math. 45, no. 2-3 (1983), 157–174.

[Waj99] B. Wajnryb, *An elementary approach to the mapping class group of a surface,* Geom. Topol. 3(1999), 405–466.

[Wall62] C.T.C. Wall, *On the orthogonal groups of unimodular quadratic forms,* Math. Ann. 147 (1962), 328–338.

[Wit94] E. Witten, *Monopoles and Four-Manifolds,* Math. Res. Lett. 1 (1994), 809–822.

Prof. Fabrizio Catanese, Lehrstuhl Mathematik VIII, Universität Bayreuth, NWII, D-95440 Bayreuth, Germany

*E-mail address*: `Fabrizio.Catanese@uni-bayreuth.de`

Prof. Michael Lönne, Mathematisches Institut der, Universität Göttingen,, Bunsenstrasse 3, D-37073 Göttingen, Germany

*E-mail address*: `Michael.Loenne@uni-bayreuth.de`

Prof. Dov Bronislaw Wajnryb, Department of Mathematics and Applied Physics, Technical University of Rzeszow, Rzeszow, Poland

*E-mail address*: `dwajnryb@prz.edu.pl`

Proceedings of $16^{th}$ Gökova
Geometry-Topology Conference
pp. 99 – 123

# Deformations of asymptotically cylindrical special Lagrangian submanifolds

*Sema Salur and A. J. Todd*

ABSTRACT. Given an asymptotically cylindrical special Lagrangian submanifold $L$ with fixed boundary in an asymptotically cylindrical Calabi-Yau 3-fold $X$, we determine conditions on a decay rate $\gamma$ which make the moduli space of (local) special Lagrangian deformations of $L$ in $X$ a smooth manifold and show that it has dimension equal to the dimension of the image of $H^1_{cs}(L,\mathbb{R})$ in $H^1(L,\mathbb{R})$ under the natural inclusion map, $[\chi] \mapsto [\chi]$. Then we prove the analogous result for asymptotically cylindrical special Lagrangian submanifolds with moving boundary.

## 1. Introduction

McLean [14] proved that the moduli space of special Lagrangian deformations of a compact special Lagrangian submanifold $L$ of a Calabi-Yau $n$-fold is a smooth manifold of dimension $b^1(L)$, the first Betti number of $L$. The goal of this paper is to prove analogous results for an asymptotically cylindrical special Lagrangian submanifold of an asymptotically cylindrical Calabi-Yau 3-fold.

Examples of complete Calabi-Yau $n$-folds are constructed by Tian and Yau in [19, 20]. In general, they construct complete Kähler metrics with a prescribed Ricci curvature on quasi-projective manifolds $M$, that is, $M = \overline{M} \setminus D$ where $\overline{M}$ is a projective manifold and $D$ a smooth, ample divisor in $\overline{M}$. Then, under further assumptions that $D$ is anticanonical and $K^{-1}_{\overline{M}}$ is ample, $K_{\overline{M}}$ the canonical line bundle of $\overline{M}$, $M$ admits a complete Ricci-flat Kähler metric. Building on this theory, Kovalev [8] constructs asymptotically cylindrical Calabi-Yau 3-folds by $W = \overline{W} \setminus D$, where $\overline{W}$ is a compact, simply-connected Kähler manifold and $D$, a $K3$ surface in $\overline{W}$, is an anticanonical divisor with trivial self-intersection class $D \cdot D = 0$ in the second integral homology class of $\overline{W}$. Given two asymptotically cylindrical Calabi-Yau 3-folds $W_1, W_2$, one can obtain asymptotically cylindrical 7-manifolds with holonomy in $SU(3)$ by taking $W_1 \times S^1$ and $W_2 \times S^1$. In his paper, Kovalev constructs examples of compact $G_2$-manifolds by gluing two such noncompact manifolds together.

We begin in Section 2 with elementary definitions from Calabi-Yau and special Lagrangian geometry then give the definitions of asymptotically cylindrical Calabi-Yau 3-folds and their asymptotically cylindrical special Lagrangian submanifolds. Included is a sketch of McLean's result in the compact case. The next section gives the material on weighted Sobolev spaces and elliptic operators on asymptotically cylindrical special

Lagrangian 3-folds developed by Lockhart and McOwen in [10, 11]. We use this theory in Section 4 to study the operator $(\mathrm{d}+\mathrm{d}^*)^p_{2+l,\gamma}$ on special Lagrangian 3-folds. In particular, we are able to determine necessary and sufficient conditions on the decay rate $\gamma$ that ensures this operator is Fredholm which then allows us to calculate its kernel and cokernel. Further, using a bit of linear algebra, we can calculate the dimension of the kernel explicitly. The final two sections prove the following results:

**Theorem 1.1.** *Assume that $(X, \omega, g, \Omega)$ is an asymptotically cylindrical Calabi-Yau 3-fold, asymptotic to $M \times S^1 \times (R, \infty)$ with decay rate $\alpha < 0$, where $M$ is a compact connected Calabi-Yau 2-fold, and that $L$ is an asymptotically cylindrical special Lagrangian 3-submanifold in $X$ asymptotic to $N \times \{p\} \times (R', \infty)$ for $R' > R$ with decay rate $\beta$ for $\alpha \leq \beta < 0$ where $N$ is a compact special Lagrangian 2-fold in $M$ and $p \in S^1$.*

*If $\gamma < 0$ is such that $\beta < \gamma$ and $(0, \gamma^2]$ contains no eigenvalues of the Laplacian on functions on $N$, then the moduli space $\mathcal{M}^\gamma_L$ of asymptotically cylindrical special Lagrangian submanifolds in $X$ near $L$, and asymptotic to $N \times \{p\} \times (R', \infty)$ with decay rate $\gamma$, is a smooth manifold of dimension $\dim V = b^2(L) - b^0(N) + b^0(L)$ where $V$ is the image of the natural inclusion map of $H^1_{cs}(L, \mathbb{R}) \hookrightarrow H^1(L, \mathbb{R})$.*

**Theorem 1.2.** *Assume that $(X, \omega, g, \Omega)$ is an asymptotically cylindrical Calabi-Yau 3-fold, asymptotic to $M \times S^1 \times (R, \infty)$ with decay rate $\alpha < 0$, where $M$ is a compact connected Calabi-Yau 2-fold, and that $L$ is an asymptotically cylindrical special Lagrangian 3-submanifold in $X$ asymptotic to $N \times \{p\} \times (R', \infty)$ for $R' > R$ with decay rate $\beta$ for $\alpha \leq \beta < 0$ where $N$ is a compact special Lagrangian 2-fold in $M$ and $p \in S^1$. Let $\gamma < 0$ be such that $\beta < \gamma$ and such that $(0, \gamma^2]$ contains no eigenvalues of the Laplacian on functions on $N$.*

*Let $\Upsilon : H^1(N, \mathbb{R}) \to H^2_{cs}(L, \mathbb{R})$ be the linear map coming from the long exact sequence in cohomology (so that $\ker \Upsilon$ is a vector subspace of $H^1(N)$), let $\mathcal{U}$ be a small open neighborhood of $0$ in $\ker \Upsilon$ and let $N_s$ denote the special Lagrangian submanifolds of $M$ near $N$ for $s \in \mathcal{U}$. Let also $N^1, \ldots, N^{b^0(N)}$ be the connected components of $N$, and let $N^i_s$ denote the special Lagrangian submanifolds of $M$ near $N^i$ for $i = 1, \ldots, b^0(N)$. Then the moduli space of asymptotically cylindrical special Lagrangian 3-folds in $X$ close to $L$, and asymptotic to $\coprod_{i=1}^{b^0(N)} N^i_s \times \{q^i\} \times (R', \infty)$ for $q^1, \ldots, q^{b^0(N)}$ in $S^1$ close to $p$, with decay rate $\gamma$, is a smooth manifold of dimension $\dim V + \dim \ker \Upsilon + b^0(N) = (b^2(L) - b^0(N) + b^0(L)) + (b^2(L) - b^1(L) + b^0(L) - b^2(N) + b^1(N)) + b^0(N)$, where $V$ is the image of $H^1_{cs}(L)$ in $H^1(L)$.*

**Remark 1.1.** In [7] and [17], similar results are proven for asymptotically cylindrical coassociative submanifolds with fixed and moving boundaries in an asymptotically cylindrical $G_2$ manifold. The arguments presented and the techniques used in this paper parallel the arguments and techniques therein.

To prove the first theorem, we construct a deformation map using the identification of the normal bundle of a special Lagrangian submanifold with its cotangent bundle. We then prove that this map extends to a smooth map of weighted Sobolev space, calculate its linearization and show that the image lies in spaces of exact forms. The map as defined will not be elliptic, so we extend it over a space of three forms to gain ellipticity; the image of this map is then shown to be in the image of the map $(\mathrm{d}+\mathrm{d}^*)^p_{2+l,\gamma}$. This allows us to use the Implicit Mapping Theorem for Banach Spaces to conclude that the kernel of our elliptic operator is smooth, finite-dimensional and locally isomorphic to the kernel of $(\mathrm{d}+\mathrm{d}^*)^p_{2+l,\gamma}$. We get the theorem then by showing that the moduli space of asymptotically cylindrical special Lagrangian deformations is homeomorphic to the kernel of our elliptic operator.

The proof of the second theorem is similar. We first prove rigidity results regarding cylindrical and asymptotically cylindrical special Lagrangians; this allows us to give precise meaning to "moving boundary at infinity." We then proceed as in the proof of Theorem 1.1; to calculate the dimension of the moduli space in this situation, we will use results obtained by Lockhart and McOwen regarding the Fredholm index of elliptic operators.

## 2. Calabi-Yau and special Lagrangian geometry

We begin with the definitions of a Calabi-Yau $n$-fold and special Lagrangian $n$-submanifold, and we give the definitions of (asymptotically) cylindrical Calabi-Yau 3-folds and (asymptotically) cylindrical special Lagrangian 3-submanifolds. References for this section include: Harvey and Lawson, [4]; Joyce, [6]; Kovalev, [8]; and McLean, [14].

**Definition 2.1.** A *complex* $n$-dimensional *Calabi-Yau* manifold $(X,\omega,g,\Omega)$ is a Kähler manifold with zero first Chern class, that is $c_1(X)=0$. In this case, $(X,\omega,g,\Omega)$ is also called a *Calabi-Yau $n$-fold.*

In [4, Section 3.1], it is proved that $\operatorname{Re}\Omega$ is a calibration form. The calibrated submanifolds with respect to $\operatorname{Re}\Omega$ are called special Lagrangian; that is, $\operatorname{Re}\Omega$ restricts to be the volume form with respect to the induced metric on special Lagrangian submanifolds. Equivalent to this condition, and what we will use as our definition, is the following:

**Definition 2.2.** A *real* $n$-dimensional submanifold $L\subseteq X$ is *special Lagrangian* if $L$ is Lagrangian (i.e. $\omega|_L\equiv 0$) and $\operatorname{Im}\Omega$ restricted to $L$ is zero. In this case, we will also call $L$ a *special Lagrangian $n$-submanifold.*

We now define cylindrical and asymptotically cylindrical Calabi-Yau 3-manifolds. See [8] for more information on the definitions in this section.

**Definition 2.3.** A Calabi-Yau 3-fold $(X_0,\omega_0,g_0,\Omega_0)$ is called *cylindrical* if $X_0$ is $M\times S^1\times\mathbb{R}$ where $M$ is a compact connected Calabi-Yau 2-fold with Kähler form $\kappa_I$ and holomorphic $(2,0)$-form $\kappa_J+i\kappa_K$, and $(\omega_0,g_0,\Omega_0)$ is compatible with the product structure $M\times S^1\times\mathbb{R}$, that is, $\omega_0=\kappa_I+\mathrm{d}\theta\wedge\mathrm{d}t$, $\Omega_0=(\kappa_J+i\kappa_K)\wedge(dt+i\mathrm{d}\theta)$ and $g_0=g_{(M\times S^1)}+\mathrm{d}t^2$.

Calabi-Yau 2-folds are also hyperkähler since they have holonomy $SU(2)=Sp(1)$. The indices on the 2-forms $\kappa_I$, $\kappa_J$ and $\kappa_K$ on $M$ are then meant to reflect the hyperkähler structure of $M$.

**Definition 2.4.** A connected, complete Calabi-Yau 3-fold $(X, \omega, g, \Omega)$ is called *asymptotically cylindrical with decay rate* $\alpha < 0$ if there exists a cylindrical Calabi-Yau 3-fold $(X_0, \omega_0, g_0, \Omega_0)$ as in Definition 2.3, a compact subset $K \subset X$, a real number $R > 0$, and a diffeomorphism $\Psi : M \times S^1 \times (R, \infty) \to X \setminus K$ such that $\Psi^*(\omega) = \omega_0 + \mathrm{d}\xi_1$ for some 1-form $\xi_1$ with $\left|\nabla^k \xi_1\right| = O(e^{\alpha t})$ and $\Psi^*(\Omega) = \Omega_0 + \mathrm{d}\xi_2$ for some complex 2-form $\xi_2$ with $\left|\nabla^k \xi_2\right| = O(e^{\alpha t})$ on $M \times S^1 \times \mathbb{R}$ for all $k \geqslant 0$, where $\nabla$ is the Levi-Cività connection of the cylindrical metric $g_0 = g_{(M \times S^1)} + dt^2$.

This definition implies that the restriction of the Kähler form $\omega$ and holomorphic $(3, 0)$-form $\Omega$ equals to the above for large $t$, up to a possible error of order $O(e^{\alpha t})$ for some parameter $\alpha < 0$. Notice also in this definition that we are assuming $X$ and $M$ are connected, so that $X$ only has one end; since we are working with Ricci-flat manifolds, this is not a restrictive assumption by a result obtained by the first author in [16].

We now define cylindrical and asymptotically cylindrical special Lagrangian submanifolds similarly.

**Definition 2.5.** Let $(X_0, \omega_0, g_0, \Omega_0)$ be a cylindrical Calabi-Yau 3-fold as in Definition 2.3. A 3-dimensional submanifold $L_0$ of $X_0$ is called *cylindrical special Lagrangian* if $L_0 = N \times \{p\} \times \mathbb{R}$ for some compact special Lagrangian submanifold $N$ in $M$, a point $p$ in $S^1$ and the restrictions of $\omega_0$ and $\operatorname{Im}\Omega_0$ to $L_0$ are zero, that is, $\omega_0|_{L_0} = (\kappa_I + \mathrm{d}\theta \wedge \mathrm{d}t)|_{L_0} = 0$ and $\operatorname{Im}(\Omega_0)|_{L_0} = \operatorname{Im}((\kappa_J + i\kappa_K) \wedge (\mathrm{d}t + i\mathrm{d}\theta))|_{L_0} = 0$.

**Definition 2.6.** Let $(X_0, \omega_0, g_0, \Omega_0)$ , $M \times S^1 \times \mathbb{R}$, $(X, \omega, g, \Omega)$, $K, R, \Psi$ and $\alpha$ be as in Definitions 2.3 and 2.4, and let $L_0 = N \times \{p\} \times \mathbb{R}$ be a cylindrical special Lagrangian 3-fold in $X_0$ as in Definition 2.5.

A connected, complete special Lagrangian 3-fold $L$ in $(X, \omega, g, \Omega)$ is called *asymptotically cylindrical with decay rate* $\beta$ for $\alpha \leq \beta < 0$ if there exists a compact subset $K' \subset L$, a normal vector field $v$ on $N \times \{p\} \times (R', \infty)$ for some $R' > R$, and a diffeomorphism $\Phi : N \times \{p\} \times (R', \infty) \to L \setminus K'$ such that the following diagram commutes:

$$\begin{array}{ccccc} M \times S^1 \times (R', \infty) & \xleftarrow[\exp_v]{} & N \times \{p\} \times (R', \infty) & \xrightarrow[\Phi]{} & (L \setminus K') \\ \big\downarrow \subset & & & & \big\downarrow \subset \\ M \times S^1 \times (R, \infty) & & \xrightarrow{\quad\quad\Psi\quad\quad} & & (X \setminus K), \end{array} \tag{1}$$

and $\left|\nabla^k v\right| = O(e^{\beta t})$ on $N \times \{p\} \times (R', \infty)$ for all $k \geq 0$.

Notice that here again, we require $L$ to be connected; however, unlike the case of the ambient manifold, it may be that $N$ is *not* connected, so that $L$ may have multiple ends. This will not present a problem in the analysis since we are assuming the ends have the same decay rate.

We now sketch a proof of McLean's result on the moduli space of special Lagrangian deformations in the case of compact special Lagrangian submanifolds of a Calabi-Yau $n$-fold.

**Theorem 2.7.** *The moduli space of deformations of a smooth, compact, orientable special Lagrangian submanifold $L$ in a Calabi-Yau manifold $X$ within the class of special Lagrangian submanifolds is a smooth manifold of dimension equal to dim($H^1(L)$).*

*Sketch of proof.* Define the deformation map $F : \nu_L \to \Lambda^2 T^*(L) \oplus \Lambda^n T^*(L)$ from the space of smooth sections of the normal bundle on $L$ to the spaces of differential 2-forms and differential $n$-forms on $L$ as follows:

$$F(V) = ((\exp_V)^*(-\omega), (\exp_V)^*(\operatorname{Im}\Omega)).$$

Note that $F$ is just the restrictions of $-\omega$ and $\operatorname{Im}\Omega$ to $L_V$ which are then pulled back to $L$ via $(\exp_V)^*$ where $\exp_V$ is the exponential map giving a diffeomorphism of $L$ onto its image $L_V$ in a neighborhood of 0.

Recall that the normal bundle $\nu_L$ of a special Lagrangian submanifold is isomorphic to the cotangent bundle $T^*(L)$. Thus there is a natural identification of normal vector fields to $L$ with differential 1-forms on $L$; moreover, since $L$ is compact, these normal vector fields can be identified with nearby submanifolds, so that, under these identifications, the kernel of $F$ corresponds to the special Lagrangian deformations.

The linearization of $F$ at 0

$$\mathrm{d}F(0) : \nu_L \to \Lambda^2 T^*(L) \oplus \Lambda^n T^*(L)$$

is given by

$$\begin{aligned} \mathrm{d}F(0)(V) =& \frac{\partial}{\partial t} F(tV)|_{t=0} = \frac{\partial}{\partial t}(\exp^*_{tV}(-\omega), \exp^*_{tV}(\operatorname{Im}\Omega))|_{t=0} \\ =& (-\mathcal{L}_V \omega|_L, \mathcal{L}_V(\operatorname{Im}\Omega)|_L). \end{aligned}$$

Using the Cartan Formula for the Lie derivative $\mathcal{L}_V$, we find

$$\begin{aligned} \mathrm{d}F(0)(V) &= (-(i_V \mathrm{d}\omega + \mathrm{d}(i_V\omega))|_L, (i_V \mathrm{d}(\operatorname{Im}\Omega) + \mathrm{d}(i_V(\operatorname{Im}\Omega)))|_L) \\ &= (-\mathrm{d}(i_V\omega)|_L, \mathrm{d}(i_V(\operatorname{Im}\Omega))|_L) = (\mathrm{d}v, \mathrm{d} * v), \end{aligned}$$

where $i_V$ represents the interior derivative and $v$ is the dual 1-form to the vector field $V$ with respect to the induced metric. Hence we have

$$\mathrm{d}F(0)(V) = (\mathrm{d}v, \mathrm{d} * v) = (\mathrm{d}v, *\mathrm{d}^* v).$$

The next step is to show that $\mathrm{d}F(0)(V) = (\mathrm{d}v, \mathrm{d} * v) = (\mathrm{d}v, *\mathrm{d}^*v)$ is onto when considered as a map from $\nu_L$ to *exact* 2-forms and *exact* $n$-forms. McLean shows this by first proving that $F$ is a map from $\nu_L$ to exact 2-forms and exact $n$-forms as follows:

Note that the image of $F$ lies in closed 2-forms and closed $n$-forms since $F$ is the pullback of the closed forms $\omega$ and $\operatorname{Im}\Omega$. Now, by replacing $V$ with $tV$, we see that $exp : L \to X$ is homotopic to the inclusion $i : L \to X$, so that $exp_V^*$ and $i^*$ induce the same map on the level of cohomology. Since $L$ is special Lagrangian, we get $[exp_V^*(\omega)] = [i^*(\omega)] = [\omega|_L] = 0$ and $[exp_V^*(\operatorname{Im}\Omega)] = [i^*(\operatorname{Im}\Omega)] = [\operatorname{Im}\Omega|_L] = 0$, so the forms in the image of $F$ are cohomologous to zero, that is, they are exact forms.

One can now easily show that, given any exact 2-form $a$ and exact $n$-form $b$, there exists a 1-form $v$ satisfying the equations $\mathrm{d}v = a$ and $\mathrm{d} * v = b$; hence, $\mathrm{d}F(0)(V)$ is surjective as claimed. Finally, after completing the spaces of differential forms with appropriate norms, we can use the Implicit Function Theorem for Banach spaces and an elliptic regularity result to conclude that $F^{-1}(0,0)$ is a smooth manifold with tangent space at 0 isomorphic to $H^1(L)$. □

## 3. Analysis on asymptotically cylindrical special Lagrangian 3-manifolds

Lockhart and McOwen [10, 11] set up a general framework for elliptic operators on noncompact manifolds where the basic tools are weighted Sobolev spaces. Using these spaces and the notion of asymptotically cylindrical linear elliptic partial differential operators, they get weighted Sobolev embedding theorems and recover elliptic regularity and Fredholm results.

**Remark 3.1.** For an alternate approach to these types of results on noncompact manifolds, see Melrose [12, 13].

Let $L$ be an asymptotically cylindrical special Lagrangian 3-manifold with data as in Definition 2.6.

**Definition 3.1.** Let $A$ be a vector bundle on $L$ with smooth metrics $h$ on the fibers and a connection $\nabla_A$ on $A$ compatible with $h$; let $A_0$ be a vector bundle on $N \times \{p\} \times \mathbb{R}$ that is invariant under translations in $\mathbb{R}$, that is a *cylindrical* vector bundle, with metrics $h_0$ on the fibers and $\nabla_{A_0}$ a connection on $A_0$ compatible with $h_0$ where $h_0$ and $\nabla_{A_0}$ are also invariant under translations in $\mathbb{R}$.

$A$, $h$ and $\nabla_A$ are said to be *asymptotically cylindrical*, asymptotic to $A_0$, $h_0$ and $\nabla_{A_0}$ respectively, if $\Phi^*(A) \cong A_0$ on $N \times \{p\} \times (R', \infty)$ with $|\Phi^*(h) - h_0| = O(e^{\beta t})$ and $|\Phi^*(\nabla_A) - \nabla_{A_0}| = O(e^{\beta t})$ as $t \to \infty$.

Recall that for $k \geq 0$,

$$L^p_k(A) = \left\{ s \in \Gamma(A) : s \text{ is } k\text{-times weakly differentiable and } \|s\|_{L^p_k} < \infty \right\},$$

where $\Gamma(A)$ is the space of *all* sections of $A$,

$$\|s\|_{L^p_k} = \left( \sum_{j=0}^{k} \int_L \left|\nabla^j_A s\right|^p \, dvol_L \right)^{1/p}$$

and $dvol_L$ is the volume element of $L$, and

$$L^p_{k,\mathrm{loc}}(A) = \{ s \in \Gamma(A) : \phi s \in L^p_k(A) \text{ for all } \phi \in C^\infty_0(L) \},$$

where $C^\infty_0(L)$ is the set of all compactly-supported smooth functions on $L$.

**Definition 3.2.** Choose a real valued smooth function $\rho : L \to \mathbb{R}$ such that $\Phi^*(\rho) \equiv t$ on $N \times \{p\} \times (R', \infty)$. By Definition 2.6, this condition prescribes $\rho$ on $L \setminus K'$, so it is only necessary to smoothly extend $\rho$ over $K'$. For $p \geq 1$, $k \geq 0$ and $\gamma \in \mathbb{R}$, the *weighted Sobolev space* $L^p_{k,\gamma}(A)$ is then the set of sections $s$ of $A$ such that $s \in L^p_{\text{loc}}(A)$, $s$ is $k$-times weakly differentiable and

$$\|s\|_{L^p_{k,\gamma}} = \left( \sum_{j=0}^{k} \int_L e^{-\gamma\rho} \left| \nabla^j_A s \right|^p dvol_L \right)^{1/p} < \infty.$$

In particular, $L^p_{k,\gamma}(A)$ is a Banach space; further, note that different choices of $\rho$ give the same space $L^p_{k,\gamma}(A)$ with equivalent norms since $\rho$ is uniquely determined outside of $K$.

Define the Banach space $C^k_\gamma(A)$ of continuous sections $s$ of $A$ with $k$ continuous derivatives such that $e^{-\gamma\rho}|\nabla^j_A s|$ is bounded for each $j = 0, 1, \ldots, k$ where the norm is given by

$$\|s\|_{C^k_\gamma} = \sum_{j=0}^{k} \sup_L e^{-\gamma\rho} \left| \nabla^j_A s \right|.$$

In general, from [2, Theorem 1.2], [10, Theorem 3.10] and [11, Lemma 7.2] there is the following Weighted Sobolev Embedding Theorem (adapted to our case):

**Theorem 3.3** (Weighted Sobolev Embedding Theorem). *Suppose that $k, l$ are integers with $k \geq l \geq 0$ and that $p, q, \gamma, \overline{\gamma}$ are real numbers with $p, q > 1$. Then*

(1) *If $\frac{k-l}{3} \geq \frac{1}{p} - \frac{1}{q}$ and $\overline{\gamma} \geq \gamma$ there is a continuous embedding of $L^p_{k,\gamma}(A) \hookrightarrow L^q_{l,\overline{\gamma}}(A)$;*
(2) *If $\frac{k-l}{3} > \frac{1}{p}$ and $\overline{\gamma} \geq \gamma$ there is a continuous embedding of $L^p_{k,\gamma}(A) \hookrightarrow C^l_{\overline{\gamma}}(A)$;*

Now suppose that $A, B$ are two asymptotically cylindrical vector bundles on $L$ which are asymptotic to the cylindrical vector bundles $A_0$ and $B_0$ on $N \times \{p\} \times \mathbb{R}$. Let $F_0 : C^\infty(A_0) \to C^\infty(B_0)$ be a cylindrical linear partial differential operator of order $k$, that is, a linear partial differential operator which is invariant under translations in $\mathbb{R}$, from *smooth* sections $C^\infty(A_0)$ of $A_0$ to *smooth* sections $C^\infty(B_0)$ of $B_0$. Let $F : C^\infty(A) \to C^\infty(B)$ be a linear partial differential operator of order $k$ on $L$.

**Definition 3.4.** $F$ is said to be an *asymptotically cylindrical operator*, asymptotic to $F_0$, if, under the identifications $\Phi^*(A) \cong A_0$, $\Phi^*(B) \cong B_0$ on $N \times \{p\} \times \mathbb{R}$, $|\Phi^*(F) - F_0| = O(e^{\beta t})$ as $t \to \infty$.

From these definitions $F$ extends to a bounded linear operator

$$F^p_{k+l,\gamma} : L^p_{k+l,\gamma}(A) \to L^p_{l,\gamma}(B)$$

for all $p > 1$, $l \geq 0$ and $\gamma \in \mathbb{R}$. We then have the following elliptic regularity result [10, Theorem 3.7.2]:

**Theorem 3.5.** *Let $F$ and $F_0$ be as above. If $1 < p < \infty$, $l \in \mathbb{Z}$ and $\gamma \in \mathbb{R}$, then for all $s \in L^p_{k+l,loc}(A)$,*

$$\|s\|_{L^p_{k+l,\gamma}} \leq C(\|Fs\|_{L^p_{l,\gamma}} + \|s\|_{L^p_{l,\gamma}})$$

*for some $C$ independent of $s$.*

**Definition 3.6.** Assume now that $F, F_0$ are also elliptic on $L$ and $N \times \{p\} \times \mathbb{R}$ respectively. Extend $F_0$ to $F_0 : C^\infty(A_0 \otimes_\mathbb{R} \mathbb{C}) \to C^\infty(B_0 \otimes_\mathbb{R} \mathbb{C})$, and define $\mathcal{D}_{F_0}$ as the set of $\gamma \in \mathbb{R}$ such that for some $\delta \in \mathbb{R}$ there exists a nonzero section $s \in C^\infty(A_0 \otimes_\mathbb{R} \mathbb{C})$, invariant under translations in $\mathbb{R}$, such that $F_0(e^{(\gamma+i\delta)t}s) = 0$.

The conditions for $F^p_{k+l,\gamma}$ to be Fredholm are now given by [11, Theorem 1.1].

**Theorem 3.7.** *Using the setup of this section, $\mathcal{D}_{F_0}$ is a discrete subset of $\mathbb{R}$, and for $p > 1$, $l \geq 0$ and $\gamma \in \mathbb{R}$, $F^p_{k+l,\gamma} : L^p_{k+l,\gamma}(A) \to L^p_{l,\gamma}(B)$ is Fredholm if and only if $\gamma \notin \mathcal{D}_{F_0}$.*

This theorem and Theorem 3.5 imply that $\ker(F^p_{k+l,\gamma})$ is a finite-dimensional vector space of smooth sections of $A$ whenever $\gamma \notin \mathcal{D}_{F_0}$, and from the Weighted Sobolev Embedding Theorem and the fact that $\ker(F^p_{k+l,\gamma})$ is invariant under small changes in $\gamma$ we have the following:

**Lemma 3.8.** *If $\gamma \notin \mathcal{D}_{F_0}$, then the kernel $\ker(F^p_{k+l,\gamma})$ is the same for any choices of $p > 1$ and $l \geq 0$ and is a finite-dimensional vector space consisting of smooth sections of $A$.*

Let $F^* : C^\infty(B) \to C^\infty(A)$ be the formal adjoint of $F$; that is, $F^*$ is the asymptotically cylindrical linear elliptic partial differential operator of order $k$ on $L$ such that

$$\langle Fs, \tilde{s}\rangle_{L^2(B)} = \langle s, F^*\tilde{s}\rangle_{L^2(A)}$$

for all compactly-supported $s \in C^\infty(A)$, $\tilde{s} \in C^\infty(B)$.

Then for $p > 1$, $l \geq 0$ and $\gamma \notin \mathcal{D}_{F_0}$, $(F^*)^q_{-l,-\gamma} : L^q_{-l,-\gamma}(B) \to L^q_{-k-l,-\gamma}(A)$ is the dual operator of $F^p_{k+l,\gamma}$ where $q > 1$ satisfies $\frac{1}{p} + \frac{1}{q} = 1$, $L^q_{-l,-\gamma}(B)$ and $L^q_{-k-l,-\gamma}(A)$ are isomorphic as Banach spaces to the dual spaces $(L^p_{l,\gamma}(B))^*$ and $(L^p_{k+l,\gamma}(A))^*$ respectively and these isomorphisms identify $(F^*)^q_{-l,-\gamma}$ with $(F^p_{k+l,\gamma})^*$. Further, since Theorem 3.5 also holds for negative differentiability, $\ker((F^*)^q_{-l,-\gamma}) = \ker((F^*)^q_{k+m,-\gamma})$ for all $m \in \mathbb{Z}$. This allows us to identify $\mathrm{coker}(F^p_{k+l,\gamma})$ with $\ker((F^*)^q_{k+m,-\gamma})^*$ for $\gamma \notin \mathcal{D}_{F_0}$, $p, q > 1$ with $\frac{1}{p} + \frac{1}{q} = 1$ and $l, m \geq 0$; moreover, the index of $F^p_{k+l,\gamma}$ is then given by

$$\mathrm{ind}(F^p_{k+l,\gamma}) = \dim\ker(F^p_{k+l,\gamma}) - \dim\ker((F^*)^q_{k+m,-\gamma}). \tag{2}$$

## 4. $\mathrm{d} + \mathrm{d}^*$ on asymptotically cylindrical special Lagrangian 3-folds

Let $L$ be an asymptotically cylindrical special Lagrangian 3-manifold with data as in Definition 2.6, and consider the asymptotically cylindrical linear elliptic operator

$$\mathrm{d} + \mathrm{d}^* : C^\infty(T^*L \oplus \Lambda^3 T^*L) \to C^\infty(\Lambda^0 T^*L \oplus \Lambda^2 T^*L),$$

with formal adjoint given by

$$\mathrm{d}^* + \mathrm{d} : C^\infty(\Lambda^0 T^*L \oplus \Lambda^2 T^*L) \to C^\infty(T^*L \oplus \Lambda^3 T^*L).$$

We study the extension

$$(\mathrm{d} + \mathrm{d}^*)^p_{2+l,\gamma} : L^p_{2+l,\gamma}(T^*L \oplus \Lambda^3 T^*L) \to L^p_{1+l,\gamma}(\Lambda^0 T^*L \oplus \Lambda^2 T^*L) \tag{3}$$

for $p > 1$, $l \geq 0$ and $\gamma \in \mathbb{R}$. Suppose further that $L$ has no compact, connected components, so that $H^3(L, \mathbb{R}) = H^0_{cs}(L, \mathbb{R}) = 0$; then $N$ is a compact, oriented 2-manifold, and $L$ is the interior of a compact, oriented 3-manifold $\overline{L}$ with boundary $\partial\overline{L} = N$.

From this we have the following long exact sequence in cohomology:

$$\begin{array}{ccccccccc} 0 \to H^0(L) \to H^0(N) \to H^1_{cs}(L) \to H^1(L) \to H^1(N) \\ \downarrow \\ 0 \leftarrow H^3_{cs}(L) \leftarrow H^2(N) \leftarrow H^2(L) \leftarrow H^2_{cs}(L) \end{array}$$

where $H^k(L) = H^k(L, \mathbb{R})$ and $H^k(N) = H^k(N, \mathbb{R})$ are the de Rham cohomology groups, $H^k_{cs}(L, \mathbb{R})$ are compactly-supported de Rham cohomology groups and $b^k(L)$, $b^k(N)$ and $b^k_{cs}(L)$ the corresponding Betti numbers. If $V \subseteq H^1(L, \mathbb{R})$ denotes the image of the natural map $H^1_{cs}(L, \mathbb{R}) \hookrightarrow H^1(L, \mathbb{R})$, $[\chi] \mapsto [\chi]$, then, from the long exact sequence, $\dim(V) = b^1_{cs}(L) - b^0(N) + b^0(L) = b^2(L) - b^0(N) + b^0(L)$.

We now summarize a number of results from the previous section applied specifically to $(\mathrm{d}+\mathrm{d}^*)^p_{2+l,\gamma}$ as well as identify the kernel and cokernel of this operator for small $\gamma < 0$.

**Theorem 4.1.** *Suppose* $\max\{\mathcal{D}_{(\mathrm{d}+\mathrm{d}^*)_0} \cap (-\infty, 0)\} < \gamma < 0$, $p, q > 1$ *with* $\frac{1}{p} + \frac{1}{q} = 1$ *and* $l$ *and* $m$ *non-negative. Then the operator* $(\mathrm{d} + \mathrm{d}^*)^p_{2+l,\gamma}$ *is Fredholm with* $\operatorname{coker}((\mathrm{d} + \mathrm{d}^*)^p_{2+l,\gamma}) \cong (\ker((\mathrm{d}^* + \mathrm{d})^q_{2+m,-\gamma}))^*$. *The kernel* $\ker((\mathrm{d} + \mathrm{d}^*)^p_{2+l,\gamma})$ *is a vector space of smooth, closed, coclosed* 1*-forms, and the map* $\ker((\mathrm{d} + \mathrm{d}^*)^p_{2+l,\gamma}) \to H^1(L, \mathbb{R})$, $\chi \mapsto [\chi]$, *induces an isomorphism of* $\ker((\mathrm{d} + \mathrm{d}^*)^p_{2+l,\gamma})$ *with the image,* $V$, *of the natural inclusion map* $H^1_{cs}(L, \mathbb{R}) \hookrightarrow H^1(L, \mathbb{R})$*; hence,* $\dim \ker((\mathrm{d} + \mathrm{d}^*)^p_{2+l,\gamma}) = \dim V$. *Finally, the kernel* $\ker(\mathrm{d}^* + \mathrm{d})^q_{2+m,-\gamma})$ *is a vector space of smooth, closed, coclosed* 0*- and* 2*-forms.*

*Proof.* Since $\gamma \notin \mathcal{D}_{(\mathrm{d}+\mathrm{d}^*)_0}$, $(\mathrm{d} + \mathrm{d}^*)^p_{2+l,\gamma}$ is Fredholm with cokernel isomorphic to $\ker((\mathrm{d}^* + \mathrm{d})^q_{2+m,-\gamma})^*$ from Lemma 3.7 and the remarks following Lemma 3.8.

For $(\eta_1, \eta_3) \in \ker((\mathrm{d} + \mathrm{d}^*)^p_{2+l,\gamma})$, $(0, 0) \equiv (\mathrm{d} + \mathrm{d}^*)(\eta_1, \eta_3) = (\mathrm{d}^*\eta_1, \mathrm{d}\eta_1 + \mathrm{d}^*\eta_3)$, from which it follows that $\eta_1$ is coclosed. $\mathrm{d}\eta_1 = -\mathrm{d}^*\eta_3$ implies that $(\mathrm{d}^*\mathrm{d} + \mathrm{d}\mathrm{d}^*)\eta_1 = 0$ and so we get $\eta_1 \in \ker((\mathrm{d}^*\mathrm{d} + \mathrm{d}\mathrm{d}^*)^p_{2+l,\gamma})$.By [7, Proposition 3.8], $\ker((\mathrm{d} + \mathrm{d}^*)^p_{2+l,\gamma}) = \ker((\mathrm{d}^*\mathrm{d}+\mathrm{d}\mathrm{d}^*)^p_{2+l,\gamma})$, and both kernels consist of smooth closed and coclosed forms. Hence $\eta_1$ is closed, and $\eta_3$ is coclosed; smoothness follows from Theorem 3.5. Using the injectivity of the map in [7, Proposition 3.9], $\eta_3 = 0$ since $H^3(L) = 0$. Thus $\ker((\mathrm{d}+\mathrm{d}^*)^p_{2+l,\gamma})$ consists of smooth, closed, coclosed 1-forms on $L$. Also, by [7, Proposition 3.9], the map $\ker((\mathrm{d} + \mathrm{d}^*)^p_{2+l,\gamma}) \to H^1(L, \mathbb{R})$ is injective with image that of the natural inclusion map $H^1_{cs}(L, \mathbb{R}) \to H^1(L, \mathbb{R})$, which is $V$ in the notation above.

Finally, let $(\eta_0, \eta_2) \in \ker((\mathrm{d}^* + \mathrm{d})^q_{2+m,-\gamma})$, so that $(0,0) = (\mathrm{d}\eta_0 + \mathrm{d}^*\eta_2, \mathrm{d}\eta_2)$ from which it immediately follows that $\eta_2$ is closed. Now, $\mathrm{d}\eta_0 = -\mathrm{d}^*\eta_2$ implies that $\eta_0, \eta_2$ lie in the kernel of $(\mathrm{d}\mathrm{d}^* + \mathrm{d}^*\mathrm{d})^q_{2+m,-\gamma}$. Again by [7, Proposition 3.8], $\ker((\mathrm{d} + \mathrm{d}^*)^q_{2+m,-\gamma}) = \ker((\mathrm{d}\mathrm{d}^* + \mathrm{d}^*\mathrm{d})^q_{2+m,-\gamma})$, and both kernels consist of smooth closed and coclosed forms. This shows both that $\eta_2$ is coclosed, and that $\eta_0$ is closed; $\eta_0$ is coclosed by definition, and smoothness, in both cases, follows from elliptic regularity (Theorem 3.5). Hence, $\ker((\mathrm{d}^* + \mathrm{d})^q_{2+m,-\gamma})$ consists of smooth, closed, coclosed 0- and 2-forms on $L$. □

**Theorem 4.2.** *Let $p > 1$, $l \geq 0$ and $\gamma \in \mathbb{R}$. Then the operator (3) is* not *Fredholm if and only if either of the following conditions hold:*

(1) $\gamma = 0$,

(2) *$\gamma^2$ is a positive eigenvalue of $\Delta = \mathrm{d}^*\mathrm{d}$ on functions on $N$.*

*Proof.* Throughout this proof let $\mathrm{d}^*_L, *_L$ operate on $L$ and $\mathrm{d}^*, *$ operate on $N$. A smooth section of $(T^*(N \times \{p\} \times \mathbb{R}) \otimes_{\mathbb{R}} \mathbb{C})$ which is invariant under translations in $\mathbb{R}$ can be written uniquely as $\eta + f_1\mathrm{d}t$ where $\eta \in C^\infty(T^*N \otimes_{\mathbb{R}} \mathbb{C})$, the smooth sections of $T^*N \otimes_{\mathbb{R}} \mathbb{C}$, and $f_1 : N \to \mathbb{C}$ is smooth. A smooth section of $\Lambda^3 T^*(N \times \{p\} \times \mathbb{R}) \otimes_{\mathbb{R}} \mathbb{C}$ invariant under translations in $\mathbb{R}$ can be written uniquely as $f_2 dvol_N \wedge \mathrm{d}t$ with $f_2 : N \to \mathbb{C}$ smooth. Then by Theorem 3.7, $(\mathrm{d} + \mathrm{d}^*_L)^p_{2+l,\gamma}$ is not Fredholm if and only if there exist $f_1, f_2, \eta$, with $f_1, f_2, \eta$ not all zero, and $\delta \in \mathbb{R}$ such that

$$(\mathrm{d} + \mathrm{d}^*_L)(e^{(\gamma+i\delta)t}(\eta + f_1\mathrm{d}t), e^{(\gamma+i\delta)t}(f_2 dvol_N \wedge \mathrm{d}t)) \equiv (0,0).$$

Thus:

$$\begin{aligned}
0 &\equiv \mathrm{d}(e^{(\gamma+i\delta)t}(\eta + f_1\mathrm{d}t)) + \mathrm{d}^*_L(e^{(\gamma+i\delta)t}(f_2 dvol_N \wedge \mathrm{d}t)) \\
&= e^{(\gamma+i\delta)t}[-(\gamma+i\delta)\eta \wedge \mathrm{d}t + \mathrm{d}\eta + \mathrm{d}f_1 \wedge \mathrm{d}t] - *_L\mathrm{d}(e^{(\gamma+i\delta)t} f_2) \\
&= e^{(\gamma+i\delta)t}[-(\gamma+i\delta)\eta \wedge \mathrm{d}t + \mathrm{d}\eta + \mathrm{d}f_1 \wedge \mathrm{d}t] - e^{(\gamma+i\delta)t} *_L [(\gamma+i\delta)f_2\mathrm{d}t + \mathrm{d}f_2] \\
&= e^{(\gamma+i\delta)t}[-(\gamma+i\delta)\eta \wedge \mathrm{d}t + \mathrm{d}\eta + \mathrm{d}f_1 \wedge \mathrm{d}t - (\gamma+i\delta)f_2 dvol_N - (*\mathrm{d}f_2) \wedge \mathrm{d}t]
\end{aligned}$$

and

$$\begin{aligned}
0 &\equiv \mathrm{d}^*_L(e^{(\gamma+i\delta)t}(\eta + f_1\mathrm{d}t)) = - *_L \mathrm{d} *_L (e^{(\gamma+i\delta)t}(\eta + f_1\mathrm{d}t)) \\
&= - *_L \mathrm{d}[e^{(\gamma+i\delta)t}(*\eta \wedge \mathrm{d}t + f_1 dvol_N)] \\
&= - *_L [(\gamma+i\delta)e^{(\gamma+i\delta)t}\mathrm{d}t \wedge f_1 dvol_N + e^{(\gamma+i\delta)t}\mathrm{d}(*\eta \wedge \mathrm{d}t)] \\
&= -e^{(\gamma+i\delta)t}[(\gamma+i\delta)f_1 + *\mathrm{d} * \eta] \\
&= -e^{(\gamma+i\delta)t}[(\gamma+i\delta)f_1 - \mathrm{d}^*\eta].
\end{aligned}$$

This yields the following system of equations in 0-, 1- and 2-forms on $N$ respectively:

$$\begin{gathered}
(\gamma+i\delta)f_1 - \mathrm{d}^*\eta \equiv 0, \\
\mathrm{d}f_1 - (\gamma+i\delta)\eta - *\mathrm{d}f_2 \equiv 0, \\
-(\gamma+i\delta)f_2 dvol_N + \mathrm{d}\eta \equiv 0.
\end{gathered}$$

From here, we immediately see that if $\gamma = 0$ then $\delta = 0$, $\eta = 0$ and $f_1 \equiv 1$, $f_2 \equiv 1$ yields a solution to the above system of equations, in which case $(\mathrm{d}+*_L\mathrm{d}_L^*)^p_{2+l,0}$ is not Fredholm. Assume now that $\gamma + i\delta \neq 0$. Since $N$ is compact, Hodge theory yields $\eta = (\eta_0, \eta_1, \eta_2)$ where $\eta_0$ is a harmonic 1-form on $N$, $\eta_1$ is an exact 1-form on $N$ and $\eta_2$ is a coexact 1-form on $N$. Splitting the above system up by harmonic, exact and coexact 1-forms on $N$ yields the following system of equations:

$$\begin{aligned}(\gamma + i\delta)\eta_0 &= 0,\\ \mathrm{d}f_1 - (\gamma + i\delta)\eta_1 &= 0,\\ - * \mathrm{d}f_2 - (\gamma + i\delta)\eta_2 &= 0,\\ (\gamma + i\delta)f_1 - \mathrm{d}^*\eta_1 &= 0,\\ (\gamma + i\delta)f_2 - *\mathrm{d}\eta_2 &= 0.\end{aligned}$$

Because we are assuming that $\gamma + i\delta \neq 0$, $\eta_0 = 0$ we first obtain:

$$\mathrm{d}f_1 = (\gamma + i\delta)\eta_1, \tag{4}$$

$$(\gamma + i\delta)f_1 = \mathrm{d}^*\eta_1. \tag{5}$$

Notice that $f_1 = 0$ if and only if $\eta_1 = 0$. Since $f_1$ and $\eta$ cannot simultaneously be zero, neither can be zero.

Taking $\mathrm{d}^*$ of (4), then substituting (5) yields:

$$\mathrm{d}^*\mathrm{d}f_1 = (\gamma + i\delta)\mathrm{d}^*\eta_1 = (\gamma + i\delta)^2 f_1,$$

so that $(\gamma + i\delta)^2$ is an eigenvalue of $\mathrm{d}^*\mathrm{d}$ on functions on $N$. Similarly, the remaining two equations

$$\begin{aligned}*\mathrm{d}f_2 - (\gamma + i\delta)\eta_2 &= 0,\\ (\gamma + i\delta)f_2 - *\mathrm{d}\eta_2 &= 0.\end{aligned}$$

also imply that $(\gamma + i\delta)^2$ is an eigenvalue of $\mathrm{d}^*\mathrm{d}$ on functions on $N$.

Since such eigenvalues must be positive, this shows that $\delta = 0$, and so $\gamma^2$ is an eigenvalue of $\mathrm{d}^*\mathrm{d}$ on functions on $N$. Conversely, assume $\Delta f = \mathrm{d}^*\mathrm{d}f = \gamma^2 f$ for some nonzero $\gamma \in \mathbb{R}$ and some smooth nonzero $f : N \to \mathbb{C}$; then $\eta = \gamma^{-1}\mathrm{d}f$ is a smooth nonzero 1-form on $N$ which satisfies the above equations.

Now, take d of (5) and substitute (4) to get:

$$\mathrm{d}\mathrm{d}^*\eta_1 = (\gamma + i\delta)\mathrm{d}f = (\gamma + i\delta)^2\eta_1.$$

Because $\eta_1$ is an exact 1-form on $N$, $(\mathrm{d}\mathrm{d}^* + \mathrm{d}^*\mathrm{d})\eta_1 = (\gamma + i\delta)^2\eta_1$ which shows that $(\gamma + i\delta)^2$ is an eigenvalue of $\mathrm{d}\mathrm{d}^* + \mathrm{d}^*\mathrm{d} = \mathrm{d}\mathrm{d}^*$ on exact 1-forms on $N$, and so $\delta = 0$ by the same reasoning as above. Hence $\gamma^2$ is an eigenvalue of $\mathrm{d}\mathrm{d}^*$ on exact 1-forms on $N$. Conversely, assume that $\mathrm{d}\mathrm{d}^*\eta = \gamma^2\eta$ for some nonzero $\gamma \in \mathbb{R}$ and some smooth nonzero exact 1-form $\eta$ on $N$; then $f = \gamma^{-1}\mathrm{d}^*\eta$ is a smooth nonzero function on $N$ which satisfies (4) and (5). Now, note that any eigenvalue of the Laplacian on exact 1-forms is an eigenvalue of the Laplacian on functions, since if $\Delta(df) = \lambda(df)$ with $\int_N f = 0$ then $\Delta f = \lambda f$ and hence we obtained the same set of eigenvalues. One could also see this from the set of equations (4)(5) which gives a bijection between the function $f_1$ and the 1-form $\eta_1$. □

## 5. Proof of Theorem 1.1

Let $(X, \omega, g_X, \Omega)$ be an asymptotically cylindrical Calabi-Yau 3-fold with decay rate $\alpha < 0$, asymptotic to the cylindrical Calabi-Yau 3-fold $(X_0, \omega_0, g_{X_0}, \Omega_0)$, $X = M \times S^1 \times \mathbb{R}$ where $M$ is a compact connected Calabi-Yau 2-fold as in Definition 2.3. Let $K \subset X$ be a compact subset, $R > 0$ and $\Psi : M \times S^1 \times (R, \infty) \to X \setminus K$ a diffeomorphism with the following properties:

(1) $\Psi^*(\omega_{X_0}) = \omega + \mathrm{d}\xi_1$, for some $\xi_1$, a 1-form on $X$ with $|\nabla^k \xi_1| = O(e^{\alpha t})$ for all $k \geq 0$;

(2) $\Psi^*(\Omega_{X_0}) = \Omega + \mathrm{d}\xi_2$, for some $\xi_2$, a complex 2-form on $X$ with $|\nabla^k \xi_2| = O(e^{\alpha t})$ for all $k \geq 0$.

Let $L$ be an asymptotically cylindrical special Lagrangian 3-submanifold of $X$ with decay rate $\beta$ ($\alpha \leq \beta < 0$), asymptotic to the cylindrical special Lagrangian 3-submanifold $L_0 = N \times \{p\} \times \mathbb{R}$ of $X_0$, where $N$ is a compact special Lagrangian 2-submanifold of $M$ and $p \in S^1$ as in Definition 2.6. Let $K' \subset L$ be a compact subset, $R' > R$, $v$ a normal vector field on $N \times \{p\} \times (R', \infty)$ with $|\nabla^k v| = O(e^{\beta t})$ for all $k \geq 0$ and $\Phi : N \times \{p\} \times (R', \infty) \to L \setminus K'$ a diffeomorphism making Diagram (1) commute.

Let $\gamma < 0$ be strictly greater than $\beta$ and be such that $(0, \gamma^2]$ does not contain eigenvalues of the Laplacian $\Delta_N = \mathrm{d}^*\mathrm{d}$ on complex-valued functions on $N$. Let $p > 3$, $l \geq 1$ and the map

$$(\mathrm{d} + \mathrm{d}^*)^p_{2+l,\gamma} : L^p_{2+l,\gamma}(T^*L \oplus \Lambda^3 T^*L) \to L^p_{1+l,\gamma}(\Lambda^0 T^*L \oplus \Lambda^2 T^*L)$$

be as in Equation (3). By Theorem 4.2, the conditions on $\gamma$ imply this operator is Fredholm, so the results of Theorem 4.1 are applicable.

We begin by constructing an identification of small sections of the normal bundle of $L$ with $X$ near $L$ that is compatible with the data on these manifolds. Let $\nu_N$ be the normal bundle of $N$ in $M$ with exponential map $\exp_N : \nu_N \to M$; let $\epsilon > 0$ such that $\exp_N : B_{2\epsilon}(\nu_N) \to T_N$ is a diffeomorphism of the subbundle $B_{2\epsilon}(\nu_N)$ whose fiber above each point is the ball of radius $2\epsilon$ about 0, with a tubular neighborhood $T_N$ of $N$ in $M$. Then $B_{2\epsilon}(\nu_N) \times \{p\} \times \mathbb{R}$ is a subbundle of the normal bundle $\nu_N \times \{p\} \times \mathbb{R}$ of $N \times \{p\} \times \mathbb{R}$ in $M \times S^1 \times \mathbb{R}$, $T_N \times \{p\} \times \mathbb{R}$ is a tubular neighborhood (also of $N \times \{p\} \times \mathbb{R}$ in $M \times S^1 \times \mathbb{R}$) and $\exp_N \times \iota \times id : B_{2\epsilon}(\nu_N) \times \{p\} \times \mathbb{R} \to T_N \times \{p\} \times \mathbb{R}$ is a diffeomorphism, where $\iota$ is the natural inclusion map and $id$ the identity map on $\mathbb{R}$. Notice that $v$ from above is a section of $\nu_N \times \{p\} \times (R', \infty)$, so since $v$ is decaying, we can assume that the graph of $v$ lies in $B_{2\epsilon}(\nu_N) \times \{p\} \times (R', \infty)$ (making $K', R'$ larger if necessary).

Let $\pi : B_\epsilon(\nu_N) \times \{p\} \times (R', \infty) \to N \times \{p\} \times (R', \infty)$ be the natural projection map. Let $\Xi : B_\epsilon(\nu_N) \times \{p\} \times (R', \infty) \to X \setminus K$ be given by $\Xi(w) = \Psi[(\exp_N \times \iota \times id)(v|_{\pi(w)} + w)]$. First, notice that since $\Xi$ is a composition of diffeomorphisms (onto their images), $\Xi$ is also a diffeomorphism (onto its image). Second, viewing $N \times \{p\} \times (R', \infty)$ as the zero section of $B_\epsilon(\nu_N) \times \{p\} \times (R', \infty)$, $\Xi|_{N \times \{p\} \times (R', \infty)} = \Psi \circ (\exp_N \times \iota \times id)|_{N \times \{p\} \times (R', \infty)} = \Psi \circ \exp_v = \Phi$ where the last two equalities follow from Definition 2.6 and the commutativity of Diagram (1). Third, $\mathrm{d}\Xi : T(B_\epsilon(\nu_N) \times \{p\} \times (R', \infty)) \to \Xi^*(T(X \setminus K)) = \Phi^*(T(X \setminus K))$ is an isomorphism.

Since $\Phi : N \times \{p\} \times (R', \infty) \to L \setminus K'$ is a diffeomorphism, the induced map $\mathrm{d}\Phi : T(N \times \{p\} \times (R', \infty))) \to T(L \setminus K')$ is an isomorphism. In fact, by the above discussion, $\mathrm{d}\Phi = \mathrm{d}\Xi|_{N\times\{p\}\times(R',\infty)} : T(B_\epsilon(\nu_N) \times \{p\} \times (R', \infty))|_{N\times\{p\}\times(R',\infty)} \to \Phi^*(T(L\setminus K'))$. Thus, define $\zeta = \mathrm{d}\Xi|_{N\times\{p\}\times(R',\infty)}$, so that

$$\begin{aligned} \zeta : \nu_N \times \{p\} \times (R', \infty) &\cong \frac{T(B_\epsilon(\nu_N) \times \{p\} \times (R', \infty))|_{N\times\{p\}\times(R',\infty)}}{T(N \times \{p\} \times (R', \infty))} \\ &\to \Phi^*(T(X \setminus K))/\Phi^*(T(L \setminus K')) \cong \Phi^*(T(X \setminus K)/T(L \setminus K')) = \Phi^*(\nu_L), \end{aligned} \tag{6}$$

where $\nu_L$ is the normal bundle of $L$; $\zeta$ is a vector bundle isomorphism of $\nu_N \times \{p\} \times (R', \infty)$ and $\Phi^*(\nu_L)$ by construction.

Finally, let $\Theta : B_{\epsilon'}(\nu_L) \to T_L$ denote an identification of $B_{\epsilon'}(\nu_L)$ (for some small $\epsilon' > 0$) with $T_L$, a tubular neighborhood of $L$ in $X$, satisfying the following properties: first, thinking of $L$ as the zero section of $B_{\epsilon'}(\nu_L)$, $\Theta|_L = id_L$; also, $\epsilon'$ should be small enough so that $\zeta^*(\Phi^*(B_{\epsilon'}(\nu_L))) \subset B_\epsilon(\nu_N) \times \{p\} \times (R', \infty) \subset \nu_N \times \{p\} \times (R', \infty)$ and $\Theta \circ \zeta = \Xi$ on $\xi^*(\Phi^*(B_{\epsilon'}(\nu_L)))$. Notice that the first condition implies that we have $\mathrm{d}\Theta|_L = id_{TL} : TL \subset T(B_{\epsilon'}(\nu_L))|_L \to TL \subset T(T_L)|_L$; the last condition uniquely defines $\Theta$ and $T_L$ on $B_{\epsilon'}(\nu_L)|_{L\setminus K'}$, so one need only smoothly extend $\Theta$ and $T_L$ to the compact subset $K'$ of $L$.

To summarize this construction, we have used the maps $\Xi$ and $\zeta$ to define $\Theta$, giving an identification of small sections of the normal bundle $\nu_L$ of $L$ with the ambient manifold $X$ near $L$, in such a way that is compatible with the diffeomorphisms $\Phi$ and $\Psi$. This allows us to identify small sections of $\nu_L$ with 3-submanifolds of $X$ near $L$ and detect the asymptotic convergence of such a submanifold to $N \times \{p\} \times (R', \infty)$ by the asymptotic convergence of small sections of $\nu_L$ to zero. Further, since $\nu_L \cong T^*L$ on the special Lagrangian submanifold $L$, we will regard $\Theta : B_{\epsilon'}(T^*L) \to T_L$.

This has the advantage that now smooth sections $\eta$ of the space $L^p_{2+l,\gamma}(B_{\epsilon'}(T^*L))$ (note that this is an open Banach subspace of the Banach manifold $L^p_{2+l,\gamma}(T^*L)$, so it is itself a Banach manifold) correspond to smooth 3-submanifolds of $X$ near $L$, and since $\eta$ is by definition a map $\eta : L \to L^p_{2+l,\gamma}(B_{\epsilon'}(T^*L))$, $\Theta \circ \eta : L \to T_L$. Hence we define our deformation map as $F : L^p_{2+l,\gamma}(B_{\epsilon'}(T^*L)) \to \Lambda^2T^*L \oplus \Lambda^3T^*L$, $F(\eta) = [(\Theta \circ \eta)^*(-\omega), (\Theta \circ \eta)^*(\mathrm{Im}\,\Omega)]$. Letting $\Gamma_\eta$ denote the graph of $\eta$ in $L^p_{2+l,\gamma}(B_{\epsilon'}(T^*L))$ and $\tilde{L} = \Theta(\Gamma_\eta)$ its image in $X$, $\tilde{L}$ is special Lagrangian precisely when $\omega|_{\tilde{L}} \equiv 0$ and $\mathrm{Im}\,\Omega|_{\tilde{L}} \equiv 0$, but this is equivalent to $F(\eta) = (0,0)$. Thus, $F^{-1}(0,0)$ parameterizes the special Lagrangian 3-submanifolds $\tilde{L}$ near $L$.

This completes our setup. Our first step now will be to prove that $F$ extends to a smooth map of Banach manifolds, and then extend $F$ over the space of three forms on $L$ $\tilde{F} = \tilde{F}^p_{2+l,\gamma} : L^p_{2+l,\gamma}(B_{\epsilon'}(T^*L) \oplus \Lambda^3T^*L) \to L^p_{1+l,\gamma}(\Lambda^2T^*L \oplus \Lambda^0T^*L)$ in order to get an elliptic operator. The linearization at 0 of this operator will be given by the operator $\mathrm{d}\tilde{F}^p_{2+l,\gamma}(0) = (\mathrm{d} + \mathrm{d}^*)^p_{2+l,\gamma}$ from above; moreover, we will show that $\tilde{F}$ actually maps into the image of $(\mathrm{d}+\mathrm{d}^*)^p_{2+l,\gamma}$. The point is to ultimately use these results to invoke the Implicit Mapping Theorem for Banach Manifolds (see, e. g., [6, Theorem 1.2.5]).

**Proposition 5.1.** *$F : L^p_{2+l,\gamma}(B_{\epsilon'}(T^*L)) \to L^p_{1+l,\gamma}(\Lambda^2 T^*L \oplus \Lambda^3 T^*L)$ is a smooth map of Banach manifolds with linearization at 0 given by $\eta \mapsto (\mathrm{d}\eta, *\mathrm{d}^*\eta)$.*

*Proof.* Begin by noting that our assumptions $p > 3$ and $l \geq 1$ yield, by the Weighted Sobolev Embedding Theorem 3.3, the continuous inclusion $L^p_{2+l,\gamma} \hookrightarrow C^1_\gamma$, so that locally $F(\eta) \in L^p_{1+l}$. Now $F(\eta)$ is simply the restriction of $\Theta^*(-\omega)$, a 2-form on $B_{\epsilon'}(T^*L)$, and of $\Theta^*(\operatorname{Im}\Omega)$, a 3-form on $B_{\epsilon'}(T^*L)$, to $\Gamma_\eta$. From the properties of $\Theta$, this means that $F(\eta)$ is equal to $\Xi^*(-\omega)$ and $\Xi^*(\operatorname{Im}\Omega)$ on $B_{\epsilon'}(\nu_N) \times \{p\} \times (R', \infty)$; further, the asymptotic properties of $\Psi$, $\Phi$ and $v$ we built into $\Xi$ imply that $\Xi^*(-\omega)$ is the sum of the pullback of the negative of the cylindrical Kähler 2-form $\omega_0$ on $M \times S^1 \times \mathbb{R}$ to $B_{\epsilon'}(\nu_N) \times \{p\} \times (R', \infty)$ and an error term which decays at a rate of $O(e^{\beta t})$; similarly, $\Xi^*(\operatorname{Im}\Omega)$ is the sum of the pullback of the 3-form $\operatorname{Im}\Omega$ on $M \times S^1 \times \mathbb{R}$ to $B_{\epsilon'}(\nu_N) \times \{p\} \times (R', \infty)$ and an error term which decays at a rate of $O(e^{\beta t})$.

Hence because $\beta < \gamma$, $F(\eta) \in L^p_{1+l,\gamma}(\Lambda^2 T^*L \oplus \Lambda^3 T^*L)$. Smoothness of $F$ follows by construction, so the first claim follows. That the linearization of $F$ at 0 is $(\mathrm{d} + *\mathrm{d}^*)^p_{2+l,\gamma}$ follows exactly as in Theorem 2.7 since the calculation is local. □

**Lemma 5.2.** *The image of $F$ lies in exact 2-forms and exact 3-forms; specifically,*

$$\begin{aligned} F(L^p_{2+l,\gamma}(B_{\epsilon'}(T^*L))) &\subset \mathrm{d}(L^p_{1+l,\gamma}(T^*L \oplus \Lambda^2 T^*L)) \\ &\subset L^p_{l,\gamma}(\Lambda^2 T^*L \oplus \Lambda^3 T^*L). \end{aligned}$$

*Proof.* Recall that $\omega$ and $\operatorname{Im}\Omega$ are closed forms, so they determine the de Rham cohomology classes $[\omega]$ and $[\operatorname{Im}\Omega]$. In particular, $\omega|_L \equiv 0$ and $\operatorname{Im}\Omega|_L \equiv 0$ since $L$ is special Lagrangian, so $[\omega|_L] = 0$ and $[\operatorname{Im}\Omega|_L] = 0$; moreover, since $T_L$, the tubular neighborhood of $L$ from above, retracts onto $L$, $[\omega|_{T_L}] = [\omega|_L]$ and $[\operatorname{Im}\Omega|_{T_L}] = [\operatorname{Im}\Omega|_L]$. Thus, there exist $\tau_1 \in C^\infty(T^*T_L)$ such that $\omega|_{T_L} = \mathrm{d}\tau_1$ and $\tau_2 \in C^\infty(\Lambda^2 T^*T_L)$ such that $\operatorname{Im}\Omega|_{T_L} = \mathrm{d}\tau_2$. Now, since $\omega|_L \equiv 0$, we can assume that $\tau_1|_L \equiv 0$; second, because $\omega$ and all its derivatives decay at a rate $O(e^{\beta t})$ to the translation-invariant 2-form $\omega_0$ on $M \times S^1 \times \mathbb{R}$, we can assume that $\tau_1$ and all its derivatives decay at a rate $O(e^{\beta t})$ to a translation invariant 1-form on $T_N \times \{p\} \times \mathbb{R}$. We can make similar assumptions regarding $\tau_2$ based on the properties of $\operatorname{Im}\Omega$.

From this, we calculate the following for $\eta \in L^p_{2+l,\gamma}(T^*L)$:

$$\begin{aligned} F(\eta) &= ((\Theta \circ \eta)^*(-\omega), (\Theta \circ \eta)^*(\operatorname{Im}\Omega)) \\ &= ((\Theta \circ \eta)^*(-\mathrm{d}\tau_1), (\Theta \circ \eta)^*(\mathrm{d}\tau_2)) \\ &= (\mathrm{d}(\Theta \circ \eta)^*(-\tau_1)), \mathrm{d}((\Theta \circ \eta)^*(\tau_2))). \end{aligned}$$

The result now follows by Proposition 5.1. □

$F$ as defined is not elliptic; we need to extend $F$ over a space of 3-forms in order to make it so. First, using the Hodge star isomorphism, we can interpret $F$ as a map $L^p_{2+l,\gamma}(B_{\epsilon'}(T^*L)) \to L^p_{1+l,\gamma}(\Lambda^2 T^*L \oplus \Lambda^0 T^*L)$. We can then make it elliptic by defining

$$\tilde{F} : L^p_{2+l,\gamma}(B_{\epsilon'}(T^*L) \oplus \Lambda^3 T^*L) \to L^p_{1+l,\gamma}(\Lambda^2 T^*L \oplus \Lambda^0 T^*L)$$

where $\tilde{F}(\eta_1, \eta_3) = F(\eta_1) + \mathrm{d}^*\eta_3$. Note that the linearization of $F$ at $\eta = 0$ is now given by $\mathrm{d}F(0)(\tau) = (\mathrm{d}\tau, \mathrm{d}^*\tau)$, so the linearization of $\tilde{F}$ at $(\eta_1, \eta_3) = (0,0)$ is given by $\mathrm{d}\tilde{F}(0,0)(\tau_1, \tau_3) = (\mathrm{d}\tau_1 + \mathrm{d}^*\tau_3, \mathrm{d}^*\tau_1)$.

**Proposition 5.3.** *Let $\mathcal{C}$ denote the image of the operator $(\mathrm{d}+\mathrm{d}^*)^p_{2+l,\gamma}$. Then*

$$\tilde{F} : L^p_{2+l,\gamma}(B_{\epsilon'}(T^*L) \oplus \Lambda^3 T^*L) \to \mathcal{C}.$$

*Proof.* By Theorem 4.1, $\operatorname{coker}((\mathrm{d}+\mathrm{d}^*)^p_{2+l,\gamma}) \cong (\ker((\mathrm{d}+\mathrm{d}^*)^q_{2+m,-\gamma}))^*$ with $\frac{1}{p} + \frac{1}{q} = 1$ and $m \geq 1$, so $\tilde{F}(\eta_1, \eta_3) = F(\eta_1) + \mathrm{d}^*\eta_3$ is in $\mathcal{C}$ if and only if

$$\langle F(\eta_1) + \mathrm{d}^*\eta_3, (\chi_0, \chi_2)\rangle_{L^2} \equiv 0 \text{ for all } (\chi_0, \chi_2) \in \ker((\mathrm{d}+\mathrm{d}^*)^q_{2+m,-\gamma}).$$

By the previous lemma $F(\eta_1) = (\mathrm{d}\tau_1, \mathrm{d}\tau_2)$ for some $\tau_1 \in L^p_{1+l,\gamma}(T^*L)$ and $\tau_2 \in L^p_{1+l,\gamma}(\Lambda^2 T^*L)$; then Theorem 4.1 implies $*\chi_0$ and $\chi_2$ are closed and coclosed 3- and 2-forms respectively which yields the following:

$$\begin{aligned}\langle F(\eta_1) + \mathrm{d}^*\eta_3, (*\chi_0, \chi_2)\rangle_{L^2} &= \langle \mathrm{d}\tau_2, *\chi_0\rangle_{L^2} + \langle \mathrm{d}\tau_1 + \mathrm{d}^*\eta_3, \chi_2\rangle_{L^2} \\ &= \langle \tau_2, \mathrm{d}^*(*\chi_0)\rangle_{L^2} + \langle \tau_1, \mathrm{d}^*\chi_2\rangle_{L^2} + \langle \eta_3, \mathrm{d}\chi_2\rangle_{L^2} = 0.\end{aligned}$$

□

The next step is to use the Implicit Mapping Theorem for Banach Spaces. Let $\mathcal{A} = \ker((\mathrm{d}+\mathrm{d}^*)^p_{2+l,\gamma})$ and $\mathcal{B}$ denote the subspace of $L^p_{2+l,\gamma}(T^*L \oplus \Lambda^3 T^*L)$ that is $L^2$-orthogonal to $\mathcal{A}$. Because $\mathcal{A}$ is finite-dimensional and the $L^2$-inner product is continuous on $L^p_{2+l,\gamma}(T^*L)$, $\mathcal{A}$ and $\mathcal{B}$ are Banach spaces such that $\mathcal{A} \oplus \mathcal{B} = L^p_{2+l,\gamma}(T^*L \oplus \Lambda^3 T^*L)$. Choose open neighborhoods $\mathcal{U}$, $\mathcal{V}$ of 0 in $\mathcal{A}$, $\mathcal{B}$, respectively, such that $\mathcal{U} \times \mathcal{V} \subset L^p_{2+l,\gamma}(B_{\epsilon'}(T^*L) \oplus \Lambda^3 T^*L)$. Then, by Proposition 5.1 and Proposition 5.3, $\tilde{F} : \mathcal{U} \times \mathcal{V} \to \mathcal{C}$ is a smooth map of Banach manifolds, $\tilde{F}(0,0) = (0,0)$ and $\mathrm{d}\tilde{F}(0,0) = (\mathrm{d}+\mathrm{d}^*)^p_{2+l,\gamma} : \mathcal{A} \oplus \mathcal{B} \to \mathcal{C}$; moreover, $(\mathrm{d}+\mathrm{d}^*)^p_{2+l,\gamma}|_{\mathcal{B}} : \mathcal{B} \to \mathcal{C}$ is an isomorphism of vector spaces by construction, and it is a homeomorphism of topological spaces by the Open Mapping Theorem. The Implicit Mapping Theorem for Banach Spaces now guarantees the existence of a connected open neighborhood $\mathcal{U}' \subset \mathcal{U}$ of 0 and a smooth function $G : \mathcal{U}' \to \mathcal{V}$ such that $G(0) = 0$ and $\tilde{F}(x) = (x, G(x)) \equiv (0,0)$ for all $x \in \mathcal{U}'$. Hence we conclude that near $(0,0)$, $\tilde{F}^{-1}(0,0) = \{(x, G(x)) : x \in \mathcal{U}'\}$, so that $\tilde{F}^{-1}(0,0)$ is smooth, finite-dimensional and locally isomorphic to $\mathcal{A} = \ker((\mathrm{d}+\mathrm{d}^*)^p_{2+l,\gamma})$.

**Lemma 5.4.** $\tilde{F}^{-1}(0,0) = F^{-1}(0,0) \times \{0\}$

*Proof.* We clearly have $F^{-1}(0,0)\times\{0\} \subseteq \tilde{F}^{-1}(0,0)$. Suppose then that $(\eta_1, \eta_3) \in \tilde{F}^{-1}(0,0)$, so $F(\eta_1) + \mathrm{d}^*\eta_3 = 0$. By Lemma 5.2, we know that $F(\eta_1) = (\mathrm{d}\tau_1, *\mathrm{d}\tau_2)$ for some $\tau_1 \in L^p_{1+l,\gamma}(T^*L)$ and $\tau_2 \in L^p_{1+l,\gamma}(\Lambda^2 T^*L)$;in particular, we must have $\mathrm{d}\tau_1 = -\mathrm{d}^*\eta_3$ and $*\mathrm{d}\tau_2 = 0$. Then

$$||\mathrm{d}^*\eta_3||^2_{L^2} = \langle \mathrm{d}^*\eta_3, \mathrm{d}^*\eta_3\rangle_{L^2} = -\langle \mathrm{d}^*\eta_3, \mathrm{d}\tau_1\rangle_{L^2} = 0.$$

Thus, $\mathrm{d}^*\eta_3 = 0$, and $\eta_3$ is constant since $\eta_3$ is a 3-form. Finally, since $\eta_3 \to 0$ near infinity in $L$, $\eta_3 = 0$ from which the claim now follows. □

The last part of the proof consists of defining a map from $F^{-1}(0,0)$ to the moduli space $M^\gamma_L$ of asymptotically cylindrical special Lagrangian deformations of $L$ near $L$; however, one technical step involved in showing the map is well defined is to show that the sections $\eta$ in $F^{-1}(0,0)$ are smooth. Theorem 1.1 will then follow from these results, Theorem 4.1 and the fact that, by the previous lemma, $F^{-1}(0,0)$ is smooth, finite-dimensional and locally isomorphic to $\ker((\mathrm{d}+\mathrm{d}^*)^p_{2+l,\gamma})$.

**Lemma 5.5.** *If $\eta \in F^{-1}(0,0)$, then $\eta \in L^p_{2+m,\gamma}(T^*L)$ for all $m \geq 1$.*

*Proof.* We begin by noting that the functional form of $F(\eta)$ is given by $H(x, \eta|_x, \nabla\eta|_x)$ where $x \in L$ and $H$ is a smooth function. Fix $m \geq 1$, and let $\nabla$ denote the Levi-Cività connection of $g_L$ on $L$. We are going to apply the Laplacian $\Delta_L = g^{ij}_L \nabla_i \nabla_j$ to $F$ which will allow us to split $F$ up in such a way that we can use a regularity result (which we will prove in the course of this argument) to increase the regularity of $\eta$.

Let $\nabla^x$ denote the derivative in the $x$-direction; let $\partial^y$ and $\partial^z$ denote the derivatives in the $y$- and $z$-directions respectively, where $y = \eta$ and $z = \nabla\eta$. Then

$$\begin{aligned}
\Delta_L(F(\eta)) &= g^{ij}_L \nabla_i \nabla_j (H(x,y,z)) \\
&= g^{ij}_L \nabla_i[(\nabla^x_j H)(x,y,z) + (\partial^y H)(x,y,z)\cdot\nabla_j\eta + (\partial^z H)(x,y,z)\cdot\nabla_j\nabla\eta] \\
&= g^{ij}_L[(\nabla^x_i\nabla^x_j H)(x,y,z) + (\nabla^x_i\partial^y H)(x,y,z)\cdot\nabla_j\eta + (\partial^y H)(x,y,z)\cdot\nabla_i\nabla_j\eta \\
&\quad +(\nabla^x_i\partial^z H)(x,y,z)\cdot\nabla_j\nabla\eta + (\partial^z H)(x,y,z)\cdot\nabla_i\nabla_j\nabla\eta \\
&\quad +(\partial^y\nabla^x_j H)(x,y,z)\cdot\nabla_i\eta + (\partial^y\partial^y H)(x,y,z)\cdot(\nabla_i\eta\otimes\nabla_j\eta) \\
&\quad +(\partial^y\partial^z H)(x,y,z)\cdot(\nabla_i\eta\otimes\nabla_j\nabla\eta) + (\partial^z\nabla^x_j H)(x,y,z)\cdot\nabla_i\nabla\eta \\
&\quad +(\partial^z\partial^y H)(x,y,z)\cdot(\nabla_i\nabla\eta\otimes\nabla_j\eta) + (\partial^z\partial^z H)(x,y,z)\cdot(\nabla_i\nabla\eta\otimes\nabla_j\nabla\eta)] \\
&= (\partial^z H)(x,y,z)\cdot\Delta_L\nabla\eta + (\partial^y H)(x,y,z)\cdot\Delta_L\eta + g^{ij}_L[(\nabla^x_i\nabla^x_j H)(x,y,z) \\
&\quad +(\nabla^x_i\partial^y H)(x,y,z)\cdot\nabla_j\eta + (\nabla^x_i\partial^z H)(x,y,z)\cdot\nabla_j\nabla\eta \\
&\quad +(\partial^y\nabla^x_j H)(x,y,z)\cdot\nabla_i\eta + (\partial^y\partial^y H)(x,y,z)\cdot(\nabla_i\eta\otimes\nabla_j\eta) \\
&\quad +(\partial^y\partial^z H)(x,y,z)\cdot(\nabla_i\eta\otimes\nabla_j\nabla\eta) + (\partial^z\nabla^x_j H)(x,y,z)\cdot\nabla_i\nabla\eta \\
&\quad +(\partial^z\partial^y H)(x,y,z)\cdot(\nabla_i\nabla\eta\otimes\nabla_j\eta) + (\partial^z\partial^z H)(x,y,z)\cdot(\nabla_i\nabla\eta\otimes\nabla_j\nabla\eta)].
\end{aligned}$$

Notice that $\Delta_L$ splits $F$ into two pieces: the only term $(\partial^z H)(x,y,z)\cdot\Delta_L\nabla\eta$ involving the third derivatives of $\eta$, and everything else which depends only on $\eta$ up to its second derivatives, which we will denote from now on by $E(x, \eta, \nabla\eta, \nabla^2\eta)$. Now, for $\eta$ fixed, let $\tilde{H}_\eta : L^p_{3+s}(T^*L) \to L^p_s(\Lambda^2 T^*L) \oplus L^p_s(\Lambda^3 T^*L)$ denote the map defined by $\sigma \mapsto \tilde{H}_\eta(\sigma) = (\partial^z H)(x, \eta|_x, \nabla\eta|_x) \cdot \Delta_L \nabla\sigma|_x$, so that $\tilde{H}_\eta$ is a third-order linear elliptic operator ($\tilde{H}_\eta$ is basically $\Delta_L \mathrm{d}$ with d the exterior derivative operator). Further, since the coefficients of $\tilde{H}_\eta$ only depend on $\eta$ and $\nabla\eta$, they are $L^p_{1+m}$ locally, and so the maximum regularity we can get from $\tilde{H}_\eta(\sigma)$ is $L^p_{1+m}$; this forces us to take $s$ such that $0 \le s \le m+1$. Of course, since the coefficients of $\tilde{H}_\eta$ are only $L^p_{1+m}$ locally (rather than smooth) we cannot use the elliptic regularity result, Theorem 3.5, so we need the following result:

**Lemma 5.6.** *Assume $\sigma \in L^p_3(T^*L)$ and $\tilde{H}_\eta(\sigma) \in L^p_m(\Lambda^2 T^*L) \oplus L^p_m(\Lambda^3 T^*L)$. Then $\sigma \in L^p_{3+m}(T^*L)$, and there exists a constant $\tilde{C} > 0$ such that*

$$\|\sigma\|_{L^p_{3+m,\gamma}} \le \tilde{C}(\|\tilde{H}_\eta(\sigma)\|_{L^p_{m,\gamma}} + \|\sigma\|_{L^p_{3,\gamma}}).$$

*Proof.* We will use results from Morrey [15, Section 6.2] to prove this lemma. If the coefficients of $\tilde{H}_\eta$ are $C^m$, then [15, Theorem 6.2.5] guarantees that $\sigma$ is locally $L^p_{3+m}$. Further, [15, Theorem 6.2.6] provides a local interior estimate of the form above where $\tilde{C} > 0$ depends on $m, p$, the domains involved, $C^m$-bounds on the coefficients of $\tilde{H}_\eta$ and a modulus of continuity for their $m$-th derivatives. Notice that if we have Hölder $C^{0,\alpha}$, $\alpha \in (0, 1)$, bounds for the $m$-th derivatives, we get our modulus of continuity; we can further simplify the problem to finding $C^{1+m,\alpha}$ bounds for $\eta$ since this will give $C^{m,\alpha}$ bounds for $\tilde{H}_\eta$ giving us the desired bounds. Recall that we are assuming $p > 3$, so that with $\alpha = 1 - p/3$, the Sobolev Embedding Theorem embeds $L^p_{2+m}$ into $C^{1+m,\alpha}$. Since $\eta \in L^p_{2+m,\gamma}$ and $\gamma < 0$, we get the desired control on $\eta$ which yields the modulus of continuity and the result. □

To finish the proof, let $\eta \in F^{-1}(0,0)$, so that $\eta \in L^p_{2+m,\gamma}(T^*L)$. From the above computation and the fact that $F(\eta) = 0$, we have $\tilde{H}_\eta(\eta) = -E(x, \eta, \nabla\eta, \nabla^2\eta)$, so that $\tilde{H}_\eta(\eta) \in L^p_{m,\gamma}(\Lambda^2 T^*L) \oplus L^p_{m,\gamma}(\Lambda^3 T^*L)$. By the regularity result above, we now have that $\eta \in L^p_{3+m,\gamma}(T^*L)$. The result now follows simply from induction on $m$. □

**Proposition 5.7.** *Let $M^\gamma_L$ be the moduli space of nearby asymptotically cylindrical special Lagrangian deformations of $L$ with decay rate $\gamma$ and asymptotic to $N \times \{p\} \times (R', \infty)$. Define $S : F^{-1}(0,0) \to \{3\text{-submanifolds of } X\}$ by $\eta \mapsto \Theta(\Gamma_\eta)$ where $\Gamma_\eta$ is the graph of $\eta$ in $B_{\epsilon'}(T^*L)$. Then $S$ is a homeomorphism of $F^{-1}(0,0)$ with a neighborhood of $L$ in $M^\gamma_L$.*

*Proof.* Let $\eta \in F^{-1}(0,0)$, let $\tilde{L} = S(\eta) \subset T_L \subset X$. By the Lemma 5.5, $\tilde{L}$ is smooth; further,

$$(0,0) = F(\eta) = ((\Theta \circ \eta)^*(-\omega), (\Theta \circ \eta)^*(\operatorname{Im}\Omega)) = (-\omega|_{\tilde{L}}, \operatorname{Im}\Omega|_{\tilde{L}}),$$

which proves $\tilde{L}$ is a special Lagrangian 3-submanifold of $X$.

We now appeal to Definition 2.6 to prove that $\tilde{L}$ is asymptotically cylindrical with decay rate $\gamma$. $\Theta \circ \eta : L \to \tilde{L}$ is a diffeomorphism, so let $\tilde{K}' = (\Theta \circ \eta)(K')$; then $\tilde{K}'$

is a compact subset of $\tilde{L}$. Let $\tilde{\Phi} = \Theta \circ \eta \circ \Phi : N \times \{p\} \times (R', \infty) \to \tilde{L} \setminus \tilde{K}'$; then $\tilde{\Phi}$ is a diffeomorphism. Finally, notice $\eta \in T^*L \cong \nu_L$, so $\zeta^* \circ \Phi^*(\eta)$ is a section of $\nu_N \times \{p\} \times (R', \infty)$, $\zeta$ defined by (6); $v$ is also a section of $\nu_N \times \{p\} \times (R', \infty)$, so we can define $\tilde{v} = v + \zeta^* \circ \Phi^*(\eta) \in \nu_N \times \{p\} \times (R', \infty)$. With this data, $\tilde{L}, \tilde{K}', \tilde{\Phi}, \tilde{v}$, Diagram (1) commutes, that is, $\Psi \circ \exp_{\tilde{v}} = \tilde{\Phi}$ on $N \times \{p\} \times (R', \infty)$.

We need finally to show that $\tilde{L}$ has the correct decay rate. First, $|\nabla^k v| = O(e^{\gamma t})$ for all $k \geq 0$ on $N \times \{p\} \times (R', \infty)$ since $L$ is asymptotically cylindrical with decay rate $\beta < \gamma$. Next, by Lemma 5.5, $\eta \in L^p_{l+2,\gamma}(T^*L)$ for all $l \geq 1$; by the Sobolev Embedding Theorem 3.3, $|\nabla^k \eta| = O(e^{\gamma \rho})$ (see Definition 3.2) for all $k \geq 0$ on $L$, so $|\nabla^k(\zeta^* \circ \Phi^*)(\eta)| = O(e^{\gamma t})$ for all $k \geq 0$ on $N \times \{p\} \times (R', \infty)$ since $\zeta$ and $\Phi$ are asymptotically cylindrical. Thus, $|\nabla^k \tilde{v}| = O(e^{\gamma t})$ on $N \times \{p\} \times (R', \infty)$ for all $k \geq 0$ on $N \times \{p\} \times (R', \infty)$, and so $\tilde{L}$ is asymptotically cylindrical special Lagrangian with decay rate $\gamma$; hence, the map $S : F^{-1}(0,0) \to M^{\gamma}_L$ is well defined.

Conversely, assume that $\tilde{L}$ is close to $L$ in $M^{\gamma}_L$, and let $\tilde{K}, \tilde{\Phi}, \tilde{v}$ be as in Definition 2.6 for $\tilde{L}$. Then there exists a unique smooth section $\eta$ of the bundle $B_{\epsilon'}(T^*L)$ with $\Theta \circ \eta : L \to \tilde{L}$ a diffeomorphism since $\tilde{L}$ and $L$ are $C^1$ close; moreover, because $\omega|_{\tilde{L}} \equiv 0$ and $\operatorname{Im} \Omega_{\tilde{L}} \equiv 0$, $F(\eta) = (0,0)$.

Now here again, as in the above construction, we have $\tilde{v} = v + \zeta^* \circ \Phi^*(\eta)$, and since $|\nabla^k \tilde{v}| = O(e^{\gamma t})$ and $|\nabla^k v| = O(e^{\gamma t})$ for all $k \geq 0$, we have $|\nabla^k \eta| = O(e^{\gamma \rho})$ for all $k \geq 0$ on $L$. Unfortunately, this estimate is only good enough to show $\eta \in L^p_{2+l,\gamma'}(B_{\epsilon'}(T^*L))$ for any $\gamma' > \gamma$; however, if $F'$ denotes $F$ with the new decay rate $\gamma'$, we now have $\eta \in F'^{-1}(0,0)$. If we further take $\gamma' > \gamma$ with $[\gamma, \gamma'] \cap \mathcal{D}_{(\mathrm{d}+\mathrm{d}^*)_0} = \emptyset$, then all of the above arguments apply also to $F'$, so that both $F, F'$ are smooth, finite-dimensional and locally isomorphic to $\ker((\mathrm{d}+\mathrm{d}^*)^p_{2+l,\gamma})$ and $\ker((\mathrm{d}+\mathrm{d}^*)^p_{2+l,\gamma'})$ respectively. Recall that these kernels depend only on the connected components of $\mathcal{D}_{(\mathrm{d}+\mathrm{d}^*)_0}$ in which the decay rates lie, so in fact these kernels are equal. Since $F^{-1}(0,0) \subseteq F'^{-1}(0,0)$, we have that $F^{-1}(0,0) = F'^{-1}(0,0)$ near 0 and, hence, $\eta \in F^{-1}(0,0) \subset L^p_{2+l,\gamma}(B_{\epsilon'}(T^*L))$.

It is left to consider the topology. Since we have identified submanifolds of $X$ with sections of the cotangent bundle $T^*L$ of $L$, we have an induced topology on $M^{\gamma}_L$ coming from some Banach norm on the sections $\eta$ of $T^*L$. Recall that $F^{-1}(0,0)$ with the topology defined by the $L^p_{2+l,\gamma}$ Banach norm is locally homeomorphic to the finite-dimensional vector space $\ker((\mathrm{d}+\mathrm{d}^*)^p_{2+l,\gamma})$. This shows that all Banach norms on the sections $\eta$ of $T^*L$ will induce the same topology on $M^{\gamma}_L$, so that $S$ is indeed a local homeomorphism. □

## 6. Proof of Theorem 1.2

Assume that $(X, \omega, g, \Omega)$, $\alpha$, $(X_0 = M \times S^1 \times \mathbb{R}, \omega_0, g_0, \Omega_0)$, $K \subset X$, $R$, $\Psi$, $L$, $\beta$, $L_0 = N \times \{p\} \times \mathbb{R}$, $K' \subset L$, $R'$, $v$, $\Phi$, $\nu_N$ and $\Theta$ are as in the proof of Theorem 1.1.

We begin with the following lemmas.

**Lemma 6.1.** *$N$ is special Lagrangian in $M$ if and only if $N \times \{p\} \times \mathbb{R}$ is special Lagrangian in $M \times S^1 \times \mathbb{R}$.*

*Proof.* Assume that $N$ is special Lagrangian in $M$. First note that for any $p \in S^1$, $\{p\} \times \mathbb{R}$ is Lagrangian in $S^1 \times \mathbb{R}$ since $(\mathrm{d}\theta \wedge \mathrm{d}t)|_{\{p\}\times\mathbb{R}} \equiv 0$. Then $\omega_0|_{N\times\{p\}\times\mathbb{R}} = (\kappa_I + \mathrm{d}\theta\wedge\mathrm{d}t)|_{N\times\{p\}\times\mathbb{R}} = \kappa_I|_N + (\mathrm{d}\theta\wedge\mathrm{d}t)|_{\{p\}\times\mathbb{R}} \equiv 0$ since $N$ is Lagrangian in $M$ and $\{p\}\times\mathbb{R}$ is Lagrangian in $S^1 \times \mathbb{R}$. Next, $\operatorname{Im}\Omega_0 = \operatorname{Im}[(\kappa_J + i\kappa_K)\wedge(\mathrm{d}t + i\mathrm{d}\theta)] = \kappa_J \wedge \mathrm{d}\theta + \kappa_K \wedge \mathrm{d}t$. Then $(\operatorname{Im}\Omega_0)|_{N\times\{p\}\times\mathbb{R}} = (\kappa_J \wedge \mathrm{d}\theta)|_{N\times\{p\}} + (\kappa_K \wedge \mathrm{d}t)|_{N\times\mathbb{R}} \equiv 0$. Thus we have that $N \times \{p\} \times \mathbb{R}$ is special Lagrangian in $M \times S^1 \times \mathbb{R}$.

Conversely, assume that $N \times \{p\} \times \mathbb{R}$ is special Lagrangian in $M \times S^1 \times \mathbb{R}$. Then $0 \equiv (\kappa_I + \mathrm{d}\theta\wedge\mathrm{d}t)|_{N\times\{p\}\times\mathbb{R}} = \kappa_I|_N$. Further, $0 \equiv (\kappa_J\wedge\mathrm{d}\theta + \kappa_K\wedge\mathrm{d}t)|_{N\times\{p\}\times\mathbb{R}} = \kappa_K|_N \wedge \mathrm{d}t|_{\mathbb{R}}$. Since $\mathrm{d}t$ is the volume form on $\mathbb{R}$, we must have $\kappa_K|_N \equiv 0$, thus proving that $N$ is special Lagrangian in $M$. □

**Lemma 6.2.** *If $L_0 = N\times\{p\}\times[a,b]$ is special Lagrangian in $X_0 = M\times S^1\times[a,b]$ for some $a, b \in \mathbb{R}$, then any special Lagrangian $\tilde{L_0}$ homologous to $L_0$ is of the form $\tilde{N} \times \{\tilde{p}\} \times [a,b]$ for some special Lagrangian submanifold $\tilde{N}$ of $M$, for some $\tilde{p} \in S^1$.*

*Proof.* Begin by normalizing the volume of $[a,b]$ to 1. By the previous result $N$ is special Lagrangian. Let $\tilde{L_0}$ be any special Lagrangian of $X$ which is homologous to $L_0$. We begin with the following calculation.

$$\begin{aligned}\operatorname{vol}(\tilde{L_0}) &= \int_{\tilde{L_0}} dvol_{\tilde{L_0}} = \int_{\tilde{L_0}} \operatorname{Re}(\Omega_0)\\ &= \int_{N\times\{p\}\times[a,b]} \operatorname{Re}(\Omega_0) = \int_{[a,b]} \operatorname{vol}(N)\mathrm{d}t = \operatorname{vol}(N)\end{aligned}$$

Now, let $\tilde{L_{0,t}} = \tilde{L_0} \cap (M \times \{\tilde{p}\} \times \{t\})$ for each $t \in [a,b]$, so that $\operatorname{vol}(\tilde{L_{0,t}}) \geq \operatorname{vol}(N)$ since $N$ is a calibrated submanifold of $M$ where equality happens if $\tilde{L_{0,t}}$ is also calibrated. Further, we always have $dvol_{\tilde{L_0}} \geq dvol_{\tilde{L_{0,t}}} \wedge \mathrm{d}t$ locally since $\tilde{L_{0,t}} \subset \tilde{L_0}$ which, by Fubini's Theorem, yields that $\operatorname{vol}(\tilde{L_0}) \geq \int_{[a,b]} \operatorname{vol}(\tilde{L_{0,t}})\mathrm{d}t$ where equality occurs precisely when $\tilde{L_0}$ is a product with $[a,b]$. Finally, we see

$$\begin{aligned}\operatorname{vol}(N) = \operatorname{vol}(\tilde{L_0}) &\geq \int_{[a,b]} \operatorname{vol}(\tilde{L_{0,t}})\mathrm{d}t\\ &\geq \int_{[a,b]} \operatorname{vol}(N)\mathrm{d}t = \operatorname{vol}(N).\end{aligned}$$

□

The next step will be to parameterize the special Lagrangian submanifolds near $N$ in $M$. Recall that $N$ is compact, so the moduli space of special Lagrangian deformations of $N$ near $N$ in $M$ can be identified with the harmonic 1-forms on $N$ which, again since $N$ is compact, can by identified with $H^1(N, \mathbb{R})$, the first de Rham cohomology group. This, however, is just a direct sum of $b^1(N)$ copies of $\mathbb{R}$, $b^1(N)$ the first betti number of $N$. For the moment then let $\mathcal{U}$ be an open, connected, simply-connected subset of $\mathbb{R}^d$ for some $d \leq b^1(N)$ containing 0. Now $N = N_0$ can be identified with $0 \in \mathcal{U}$, and we can talk about the special Lagrangian submanifolds $N_s$ close to $N_0$ for $s$ close to 0 in $\mathcal{U}$. We want to restrict $\mathcal{U}$ in such a way that the elements of $\mathcal{U}$ correspond to the special Lagrangian submanifolds $N_s$ in $M$ which are a boundary at infinity of an asymptotically cylindrical special Lagrangian deformation of $L$.

Recall that since $N$ is special Lagrangian there is an identification of a tubular neighborhood $T_N$ of $N$ with 1-forms on $N$. Identifying $N$ with the zero section of $T^*N$, we know that $N_s$ is identified with the graph of some harmonic 1-form $\xi_s$ on $N$; in fact, we may realize $\xi_s$ as a translation-invariant, harmonic 1-form on $N \times \{p\} \times (R', \infty)$, so the graph of $\xi_s$ corresponds to $N_s \times \{p\} \times (R', \infty)$. We then get a 1-form $\sigma_s$ on $L$ via the diffeomorphism $\Phi : N \times \{p\} \times (R', \infty) \to L \setminus K'$ ($\Phi$ gives $\sigma_s$ on $L \setminus K'$, so smoothly extend $\sigma_s$ over the compact set $K'$). Finally, the map $\Theta$ identifies the graph of $\sigma_s$ with a 3-submanifold $L_s$ of $X$ asymptotic to $N_s \times \{p\} \times \mathbb{R}$.

Notice that for each $s$ we have chosen a section $\sigma_s$, so that we need to show the independence of this choice on the problem. What we are interested in are the values of $\omega$ and $\operatorname{Im}\Omega$ when restricted to the submanifolds $L_s$, so we need only show independence of the choice of $L_s$ on these values. Note that when $L_s$ is asymptotic to $N_s \times \{p\} \times (R', \infty)$, then near infinity $\omega|_{L_s} = O(e^{\gamma t})$ and $(\operatorname{Im}\Omega)|_{L_s} = O(e^{\gamma t})$. (We will prove this fact in Proposition 6.3.) On the other hand, since $N_s$ is a special Lagrangian, we can take $\omega = 0$, $\operatorname{Im}\Omega = 0$ in a small neighborhood of the boundary $N_s$ in $N_s \times \{p\} \times (R', \infty)$, so that $\omega|_{L_s}$, $(\operatorname{Im}\Omega)|_{L_s}$ are compactly supported; i.e., we can take $[\omega|_{L_s}]$ in $H^2_{cs}(L_s, \mathbb{R})$, and $[(\operatorname{Im}\Omega)|_{L_s}]$ in $H^3_{cs}(L_s, \mathbb{R})$. Moreover, since $N_0$ and $N_s$ are isotopic we can take $[\omega|_{L_s}] \in H^2_{cs}(L_0, \mathbb{R})$ and $[(\operatorname{Im}\Omega)|_{L_s}] \in H^3_{cs}(L_0, \mathbb{R})$ which are independent of the choice of $L_s$; by Lemma 6.2, we in fact have that each $L_s$ is asymptotically cylindrical special Lagrangian, asymptotic to $N_s \times \{p\} \times \mathbb{R}$.

From topology, we have a short exact sequence of differential forms:

$$0 \to \Lambda^*_{cs}(T^*L) \to \Lambda^*(T^*L) \to \Lambda^*(T^*N) \to 0.$$

Let $i : \Lambda^*_{cs}(T^*L) \to \Lambda^*(T^*L)$ and $\varrho : \Lambda^*(T^*L) \to \Lambda^*(T^*N)$ be the corresponding chain maps. This yields a long exact sequence in de Rham cohomology given by the zig-zag lemma:

$$\begin{array}{c} 0 \to H^0(L) \to H^0(N) \to H^1_{cs}(L) \to H^1(L) \to H^1(N) \\ \downarrow \\ 0 \leftarrow H^3_{cs}(L) \leftarrow H^2(N) \leftarrow H^2(L) \leftarrow H^2_{cs}(L) \end{array} \qquad (7)$$

In particular, consider the boundary map $\Upsilon : H^1(N) \to H^2_{cs}(L)$. This map is defined as follows: for a closed 1-form $\eta$ on $N$, there is a 1-form $\tilde{\eta}$ on $L$ by the short exact sequence, such that $\varrho(\tilde{\eta}) = \eta$. Notice that $\varrho(\mathrm{d}\tilde{\eta}) = \mathrm{d}\varrho(\tilde{\eta}) = \mathrm{d}\eta = 0$. Again, by the short exact sequence above, there then exists $\widehat{\eta}$, a compactly-supported 2-form on $L$, such that $i(\widehat{\eta}) = \mathrm{d}\tilde{\eta}$. Noting that $i(\mathrm{d}\widehat{\eta}) = \mathrm{d}i(\widehat{\eta}) = \mathrm{d}\mathrm{d}\tilde{\eta} = 0$, we have $\mathrm{d}\widehat{\eta} = 0$ since $i$ is injective, that is, $\widehat{\eta}$ is a closed, compactly-supported 2-form on $L$. Thus, we define $\Upsilon([\eta]) = [\widehat{\eta}]$. From here, we have that $[\widehat{\eta}] = 0$ if and only if $\widehat{\eta} = \mathrm{d}\overline{\eta}$ for a compactly-supported 1-form $\overline{\eta}$ on $L$. By Theorem 1.1, $\overline{\eta}$ corresponds to an asymptotically cylindrical special Lagrangian

3-fold near $L$ in $X$. Thus, we define $\mathcal{U}$ to be an open, connected, simply-connected subset of $\ker \Upsilon$.

Recall from the proof of Theorem 1.1 we had the map $\Theta : B_{\epsilon'}(T^*L) \to T_L \subset X$, simply the explicit identification of small sections of $T^*L$ with the tubular neighborhood $T_L$ which is compatible with our asymptotic identifications. Now, we can assume without loss of generality that for $\eta \in L^p_{2+l,\gamma}(B_{\epsilon'}(T^*L))$ and $\sigma_s$ defined above we have a smooth section of the bundle $B_{\epsilon'}(T^*L)$ given by $\eta + \sigma_s : L \to B_{\epsilon'}(T^*L)$, so that $\Theta \circ (\eta + \sigma_s) : L \to T_L$ makes sense. Thus we can define our deformation map as

$$G : \mathcal{U} \times L^p_{2+l,\gamma}(B_{\epsilon'}(T^*L)) \to \{\text{2-forms on } L\} \oplus \{\text{3-forms on } L\}$$

by

$$G(s,\eta) = ((\Theta \circ (\eta + \sigma_s))^*(-\omega), (\Theta \circ (\eta + \sigma_s))^*(\operatorname{Im}\Omega)).$$

Note that this map is equivalent to the one defined by

$$G(s,\eta) = (\pi_*(-\omega|_{\Gamma(\eta+\sigma_s)}), \pi_*((\operatorname{Im}\Omega)|_{\Gamma(\eta+\sigma_s)}))$$

where $\pi$ is the projection onto $L$ and $\Gamma(\eta + \sigma_s)$ is the graph of $\Gamma(\eta + \sigma_s)$ in $X$, that is, a 3-submanifold $\tilde{L}$ of $X$. Finally, notice that $\tilde{L}$ is special Lagrangian if and only if $G(s,\eta) = (0,0)$ so that $G^{-1}(0,0)$ parameterizes the special Lagrangian submanifolds near $L$ in $X$.

We now need to show that $G^{-1}(0,0)$ is smooth, finite-dimensional and locally isomorphic to $\ker((\mathrm{d} + \mathrm{d}^*)^p_{2+l,\gamma}) \subset \mathcal{U} \times L^p_{2+l,\gamma}(T^*L)$. This is accomplished in much that same way as in the fixed boundary case. We show that $G$ is a smooth map when considered as a map between weighted Sobolev spaces, extend $G$ to an elliptic operator $\tilde{G}$, prove that the image of $\tilde{G}$ lies in the image of the asymptotically cylindrical linear elliptic operator $(\mathrm{d} + \mathrm{d}^*)^p_{2+l,\gamma}$ which will then allow us to use the Implicit Mapping Theorem for Banach Manifolds.

**Proposition 6.3.** *$G : \mathcal{U} \times L^p_{2+l,\gamma}(B_{\epsilon'}(T^*L)) \to L^p_{1+l,\gamma}(\Lambda^2 T^*L) \oplus L^p_{1+l,\gamma}(\Lambda^3 T^*L)$ is a smooth map of Banach manifolds with linearization at 0 given by $\mathrm{d}G_{(0,0)}(s,\eta) = (\mathrm{d}(\eta + \sigma_s), *\mathrm{d}^*(\eta + \sigma_s))$.*

*Proof.* The functional form of $G$ is

$$G(s,\eta)|_x = H(s, x, \eta|_x, \nabla\eta|_x), \; x \in L$$

where $H$ is a smooth function of its arguments. Since $p > 3$ and $l \geq 1$ by Sobolev embedding theorem we have $L^p_{2+l,\gamma}(T^*L) \hookrightarrow C^1_\gamma(T^*L)$. General arguments then show that locally $G(s,\eta)$ is $L^p_{1+l}$.

From Proposition 5.1, we know that $G(0,\eta)$ is in $L^p_{1+l,\gamma}(\Lambda^2 T^*L) \oplus L^p_{1+l,\gamma}(\Lambda^3 T^*L)$. When $s \neq 0$, then $G(s,0) = (\pi_*(-\omega|_{\Gamma(\sigma_s)}), \pi_*((\operatorname{Im}\Omega)|_{\Gamma(\sigma_s)}))$. By construction $\Gamma(\sigma_s)$ is asymptotic to $N_s \times \{p\} \times \mathbb{R}$ which is special Lagrangian in $M \times S^1 \times \mathbb{R}$ as $N_s$ is a special Lagrangian 2-fold. Then $\omega$ and $\operatorname{Im}\Omega$ on $X \setminus K$ can be written as $\omega = \kappa_I + \mathrm{d}\theta \wedge \mathrm{d}t + O(e^{\alpha t})$ and $\operatorname{Im}\Omega = \operatorname{Im}[(\kappa_J + i\kappa_K) \wedge (\mathrm{d}\theta + i\mathrm{d}t)] + O(e^{\alpha t})$ where $\alpha$ is the rate for $X$ converging to $M \times S^1 \times \mathbb{R}$. $\Gamma(\sigma_s)$ is the graph of $\sigma_s$ which is equal to $N_s \times \{p\} \times (R+1, \infty)$

in $M \times S^1 \times (R+1,\infty)$. Since $N_s$ is special Lagrangian, for $t > R+1$, $\omega|_{\Gamma(\sigma_s)} = (\kappa_I + \mathrm{d}\theta \wedge \mathrm{d}t + O(e^{\alpha t}))|_{N_s \times \{p\} \times (R+1,\infty)} = 0 + O(e^{\alpha t})|_{N_s \times \{p\} \times (R+1,\infty)}$ and $(\mathrm{Im}\,\Omega)|_{\Gamma(\sigma_s)} = (\mathrm{Im}[(\kappa_J + i\kappa_K) \wedge (\mathrm{d}\theta + i\mathrm{d}t)] + O(e^{\alpha t}))|_{N_s \times \{p\} \times (R+1,\infty)} = 0 + O(e^{\alpha t})|_{N_s \times \{p\} \times (R+1,\infty)}$.

Therefore for $\Gamma(\sigma_s)$ the error terms $\omega|_{\Gamma(\sigma_s)}$ and $(\mathrm{Im}\,\Omega)|_{\Gamma(\sigma_s)}$ come from the degree of the asymptotic decay. In particular, as $\alpha < \gamma$ we can assume $\omega|_{\Gamma(\sigma_s)} \equiv O(e^{\gamma t})$ and $(\mathrm{Im}\,\Omega)|_{\Gamma(\sigma_s)} \equiv O(e^{\gamma t})$. We can easily arrange to choose $\sigma_s$ such that $\pi_*(\omega|_{\Gamma(\sigma_s)})$ lies in $L^p_{1+l,\gamma}(\Lambda^2 T^* L)$ and $\pi_*((\mathrm{Im}\,\Omega)|_{\Gamma(\sigma_s)}) \in L^p_{1+l,\gamma}(\Lambda^3 T^* L)$ and that we have inequalities $\|\pi_*(\omega|_{\Gamma(\sigma_s)})\|_{L^p_{1+l,\gamma}} \leq c_1|s|$ and $\|\pi_*((\mathrm{Im}\,\Omega)|_{\Gamma(\sigma_s)})\|_{L^p_{1+l,\gamma}} \leq c_2|s|$ for some constants $c_1$ and $c_2$. This implies that $G(s,\eta) \in L^p_{1+l,\gamma}(\Lambda^2 T^* L) \oplus L^p_{1+l,\gamma}(\Lambda^3 T^* L)$.

The linearization of $G$ is $\mathrm{d} + *\mathrm{d}^*$ by the same calculation as before since we are still only working locally. □

**Proposition 6.4.** *The image of $G$ lies in exact 2-forms and exact 3-forms; specifically,*

$$G(\mathcal{U} \times L^p_{2+l,\gamma}(B_{\epsilon'}(T^* L))) \subset \mathrm{d}(L^p_{1+l,\gamma}(T^* L)) \oplus \mathrm{d}(L^p_{1+l,\gamma}(\Lambda^2 T^* L))$$
$$\subset L^p_{l,\gamma}(\Lambda^2 T^* L) \oplus L^p_{l,\gamma}(\Lambda^3 T^* L).$$

*Proof.* We need only modify slightly the proof from Lemma 5.2. Recall that $\omega$ and $\mathrm{Im}\,\Omega$ are closed forms, so they determine the de Rham cohomology classes $[\omega]$ and $[\mathrm{Im}\,\Omega]$. In particular, $\omega|_L \equiv 0$ and $\mathrm{Im}\,\Omega|_L \equiv 0$ since $L$ is special Lagrangian, so $[\omega|_L] = 0$ and $[\mathrm{Im}\,\Omega|_L] = 0$; moreover, since $T_L$, the tubular neighborhood of $L$ from above, retracts onto $L$, $[\omega|_{T_L}] = [\omega|_L]$ and $[\mathrm{Im}\,\Omega|_{T_L}] = [\mathrm{Im}\,\Omega|_L]$ in $H^2(T_L)$ and $H^3(T_L)$. Thus, there exists $\tau_1 \in C^\infty(T^* T_L)$ such that $\omega|_{T_L} = \mathrm{d}\tau_1$ and $\tau_2 \in C^\infty(\Lambda^2 T^* T_L)$ such that $\mathrm{Im}\,\Omega|_{T_L} = \mathrm{d}\tau_2$. Now, since $\omega|_L \equiv 0$, we can assume that $\tau_1|_L \equiv 0$; second, because $\omega$ and all its derivatives decay at a rate $O(e^{\beta t})$ to the translation-invariant 2-form $\omega_0$ on $M \times S^1 \times \mathbb{R}$, we can assume that $\tau_1$ and all its derivatives decay at a rate $O(e^{\beta t})$ to a translation invariant 1-form on $T_N \times \{p\} \times \mathbb{R}$. We can make similar assumptions regarding $\tau_2$ based on the properties of $\mathrm{Im}\,\Omega$. In particular, since $\beta < \gamma$, we have that these forms and all of their derivatives decay at $O(e^{\gamma t})$. Thus, we have $\tau_1 \in L^p_{1+l,\gamma}(T^* L)$ and $\tau_2 \in L^p_{1+l,\gamma}(\Lambda^2 T^* L)$.

By Proposition 6.3, the map $G$ maps $\mathcal{U} \times L^p_{2+l,\gamma}(B_{\epsilon'}(T^* L)) \to L^p_{1+l,\gamma}(\Lambda^2 T^* L) \oplus L^p_{1+l,\gamma}(\Lambda^3 T^* L)$. Thus for $(s,\eta) \in \mathcal{U} \times L^p_{2+l,\gamma}(T^* L)$ we have:

$$\begin{aligned} G(s,\eta) &= ((\Theta \circ (\eta + \sigma_s))^*(-\omega), (\Theta \circ (\eta + \sigma_s))^*(\mathrm{Im}\,\Omega)) \\ &= ((\Theta \circ (\eta + \sigma_s))^*(-\mathrm{d}\tau_1), (\Theta \circ (\eta + \sigma_s))^*(\mathrm{d}\tau_2)) \\ &= (\mathrm{d}(\Theta \circ (\eta + \sigma_s))^*(-\tau_1), \mathrm{d}(\Theta \circ (\eta + \sigma_s))^*(\tau_2)). \end{aligned}$$

□

We now extend the map $G$ over the space of 3-forms on $L$ to get an elliptic operator $\tilde{G}$ as we did in the proof of Theorem 1.1. Using the Hodge star isomorphism, we can view $G : \mathcal{U} \times L^p_{2+l,\gamma}(B_{\epsilon'}(T^* L)) \to L^p_{1+l,\gamma}(\Lambda^2 T^* L \oplus \Lambda^0 T^* L)$, then define $\tilde{G} : \mathcal{U} \times L^p_{2+l,\gamma}(B_{\epsilon'}(T^* L) \oplus \Lambda^3 T^* L) \to L^p_{1+l,\gamma}(\Lambda^2 T^* L \oplus \Lambda^0 T^* L)$ by $\tilde{G}(s,\eta_1,\eta_3) =$

$G(s,\eta_1)+\mathrm{d}^*\eta_3$. In this way, $\mathrm{d}G(0,0)(s,\eta_1)=(\mathrm{d}(\eta_1+\sigma_s),\mathrm{d}^*(\eta_1+\sigma_s))$ by Proposition 6.3, so $\mathrm{d}\tilde{G}(0,0)(s,\eta_1,\eta_3)=(\mathrm{d}(\eta_1+\sigma_s)+\mathrm{d}^*\eta_3,\mathrm{d}^*(\eta_1+\sigma_s))$

**Proposition 6.5.** *Let $\mathcal{C}$ denote the image of the operator $(\mathrm{d}+\mathrm{d}^*)^p_{2+l,\gamma}$. Then*

$$\tilde{G}:\mathcal{U}\times L^p_{2+l,\gamma}(B_{\epsilon'}(T^*L)\oplus\Lambda^3T^*L)\to\mathcal{C}.$$

*Proof.* Let $(s,\eta_1,\eta_3)\in\mathcal{U}\times L^p_{2+l,\gamma}(B_{\epsilon'}(T^*L)\oplus\Lambda^3T^*L)$. From Theorem 4.1, we have $\operatorname{coker}((\mathrm{d}+\mathrm{d}^*)^p_{2+l,\gamma})\cong(\ker((\mathrm{d}+\mathrm{d}*)^q_{2+m,-\gamma}))^*$ with $\frac{1}{p}+\frac{1}{q}=1$ and $m\geq 1$, so $\tilde{G}(s,\eta_1,\eta_3)=G(s,\eta_1)+\mathrm{d}^*\eta_3\in\mathcal{C}$ if and only if

$$\langle G(s,\eta_1)+\mathrm{d}^*\eta_3,(\chi_0,\chi_2)\rangle_{L^2}\equiv 0 \text{ for all } (\chi_0,\chi_2)\in\ker((\mathrm{d}+\mathrm{d}*)^q_{2+m,-\gamma}).$$

By the previous proposition $G(s,\eta_1)=(\mathrm{d}\tau_1,\mathrm{d}\tau_2)$ for some $\tau_1\in L^p_{1+l,\gamma}(T^*L)$ and $\tau_2\in L^p_{1+l,\gamma}(\Lambda^2T^*L)$; then Theorem 4.1 shows $*\chi_0$ and $\chi_2$ are harmonic 3- and 2-forms respectively which yields the following:

$$\begin{aligned}\langle G(s,\eta_1)+\mathrm{d}^*\eta_3,(*\chi_0,\chi_2)\rangle_{L^2}&=\langle\mathrm{d}\tau_1+\mathrm{d}^*\eta_3,\chi_2\rangle_{L^2}+\langle\mathrm{d}\tau_2,*\chi_0\rangle_{L^2}\\&=\langle\tau_1,\mathrm{d}^*\chi_2\rangle_{L^2}+\langle\eta_3,\mathrm{d}\chi_2\rangle_{L^2}+\langle\tau_2,\mathrm{d}^*(*\chi_0)\rangle_{L^2}=0.\end{aligned}$$

□

We are now able to invoke to the Implicit Mapping Theorem for Banach Spaces to conclude that $\tilde{G}^{-1}(0,0)$ is smooth, finite-dimensional and locally isomorphic to $\ker((\mathrm{d}+\mathrm{d}^*)^p_{2+l,\gamma})\subset\mathcal{U}\times L^p_{2+l,\gamma}(T^*L)$ as in the proof of Theorem 1.1. By a completely analogous proof to the that of Lemma 5.4, we have $\tilde{G}^{-1}(0,0)=G^{-1}(0,0)\times\{0\}$.

The final step is to calculate the dimension of the moduli space which we do by using Fredholm index arguments using the elliptic operator $\tilde{G}$. The rest of the proof of Theorem 1.2 then follows exactly as in the proof of Theorem 1.1, so we will not repeat it here.

**Proposition 6.6.** *Let $\mathcal{U}\subset H^1(N,\mathbb{R})$ be a subspace of special Lagrangian deformations of the boundary $N$. Also, let $\mathrm{d}\tilde{G}_{(0,0)}(s,\eta)$ represent the linearization of the deformation map $G$ at 0 with moving boundary and $\mathrm{d}\tilde{G}^f_{(0,0)}(\eta)$ represent the linearization of the deformation map $\tilde{G}^f$ at 0 with fixed boundary. Then*

$$\operatorname{Ind}(\mathrm{d}\tilde{G}_{(0,0)})=\dim\mathcal{U}+\operatorname{Ind}(\mathrm{d}\tilde{G}^f_{(0,0)}),$$

*where* Ind *denotes the index of a map.*

*Proof.* At $s=0$

$$\tilde{G}(0,\eta)=\tilde{G}^f(\eta)=(\pi_*(-\omega|_{\Gamma_\eta}),*\pi_*((\operatorname{Im}\Omega)|_{\Gamma_\eta}))$$

with linearization at $(0,0)$

$$\mathrm{d}\tilde{G}_{(0,0)}(0,\eta)=\mathrm{d}\tilde{G}^f_{(0,0)}(\eta).$$

Since $\mathrm{d}\tilde{G}_{(0,0)}$ is linear, we have

$$\mathrm{d}\tilde{G}_{(0,0)}(s,\eta)=\mathrm{d}\tilde{G}_{(0,0)}(s,0)+\mathrm{d}\tilde{G}^f_{(0,0)}(0,\eta)$$

where $d\tilde{G}_{(0,0)}(s,0)$ is finite-dimensional, $s \in T_0\mathcal{U} \cong \mathbb{R}^d$, $d = \dim\ker\Upsilon$. Then

$$\begin{aligned}\operatorname{Ind}(d\tilde{G}_{(0,0)} &: \mathcal{U} \times L^p_{2+l,\gamma}(T^*L \oplus \Lambda^3T^*L) \to L^p_{1+l,\gamma}(\Lambda^2T^*L \oplus \Lambda^0T^*L)) \\ = \operatorname{Ind}(d\tilde{G}^f_{(0,0)} &: \mathcal{U} \times L^p_{2+l,\gamma}(T^*L \oplus \Lambda^3T^*L) \to L^p_{1+l,\gamma}(\Lambda^2T^*L \oplus \Lambda^0T^*L)) \\ &= \dim\mathcal{U} + \operatorname{Ind}(d\tilde{G}^f_{(0,0)}).\end{aligned}$$

□

**Proposition 6.7.** *The dimension of* $\mathcal{M}^\gamma_L$*, the moduli space of special Lagrangian deformations of an asymptotically cylindrical special Lagrangian submanifold* $L$ *asymptotic to* $N_s \times \{p\} \times (R,\infty)$*,* $s \in \mathcal{U}$*, with decay rate* $\gamma$ *is*

$$\dim\mathcal{M}^\gamma_L = \dim V + b^2(L) - b^1(L) + b^0(L) - b^2(N) + b^1(N).$$

*If we further allow* $p \in S^1$ *to vary, we get a total dimension equal to* $\dim\mathcal{M}^\gamma_L + b^0(N)$.

*Proof.* Let $b^k(N)$, $b^k_{cs}(L)$ and $b^k(L)$ be the corresponding Betti numbers. We know from the previous proposition that

$$\operatorname{Ind}(d\tilde{G}_{(0,0)}) = \dim\mathcal{U} + \operatorname{Ind}(d\tilde{G}^f_{(0,0)}).$$

In particular,

$$\dim(\ker d\tilde{G}_{(0,0)}) = \dim(\ker\Upsilon) + \dim(\ker d\tilde{G}^f_{(0,0)}),$$

that is, the dimension of the moduli space for the moving boundary case is the sum of the dimension of the moduli space for the fixed boundary case and the dimension of the kernel of $\Upsilon$. Taking alternating sums of dimensions in the long exact sequence (7) shows that the dimension of the kernel of $\Upsilon$ is

$$\begin{aligned}\dim(\ker\Upsilon) &= b^1(N) - b^2_{cs}(L) + b^2(L) - b^2(N) + b^3_{cs}(L) \\ &= b^2(L) - b^1(L) + b^0(L) - b^2(N) + b^1(N).\end{aligned}$$

Finally, we show that the moduli space $\mathfrak{M}$ of asymptotically cylindrical special Lagrangian submanifolds $\tilde{L}$, asymptotic to $N_s \times \{\tilde{p}\} \times \mathbb{R}$, which are isotopic to $L$ is equivalent to the product of $S^1$ with the moduli space $\mathcal{M}^\gamma_L$ of asymptotically cylindrical special Lagrangian submanifolds $\overline{L}$, asymptotic to $N_s \times \{p\} \times \mathbb{R}$, which are isotopic to $L$, i.e., the total moduli space of asymptotically cylindrical special Lagrangian submanifolds of $X$ isotopic to $L$ is a smooth manifold of dimension $\dim(\mathfrak{M}) = \dim(\mathcal{M}^\gamma_L) + b^0(N)$.

Note that for any $\tilde{p} \in S^1$, the moduli space of asymptotically cylindrical special Lagrangians, asymptotic to $N_s \times \{\tilde{p}\} \times \mathbb{R}$, isotopic to $L$, is equivalent to the moduli space of asymptotically cylindrical special Lagrangians, asymptotic to $N_s \times \{p\} \times \mathbb{R}$. To see this, let $\tilde{L}$ be asymptotic to $N_s \times \{\tilde{p}\} \times \mathbb{R}$; then there exists an isotopy from $N_s \times \{\tilde{p}\} \times \mathbb{R}$ to $N_s \times \{p\} \times \mathbb{R}$. By the data of an asymptotically cylindrical special Lagrangian, this gives an isotopy on $\tilde{L}$ via the pullback map. Since $\tilde{L}$ is isotopic to $L$, we have an asymptotically cylindrical special Lagrangian submanifold, asymptotic to $N_s \times \{p\} \times \mathbb{R}$ and isotopic to $L$. Conversely, if $\tilde{L}$ is asymptotic to $N_s \times \{p\} \times \mathbb{R}$, there is an isotopy from $N_s \times \{p\} \times \mathbb{R}$ to $N_s \times \{\tilde{p}\} \times \mathbb{R}$, giving an isotopy on $\tilde{L}$, and hence, an asymptotically cylindrical special Lagrangian submanifold, asymptotic to $N_s \times \{\tilde{p}\} \times \mathbb{R}$ and isotopic to $L$. Finally, if $N_1,...,N_k$ are the connected components of $N$, $k = b^0(N)$, then the dimension of the moduli space of asymptotically cylindrical special Lagrangians limiting to the union of $N_i \times p_i$ for $p_1, ..., p_k$ in $S^1$ will have an additional $b^0(N)$ as we allow $p_i$ to vary. □

## References


[1] Akbulut, S. and Salur, S., *Calibrated Manifolds and Gauge Theory*, J. Reine Angew. Math., **625** (2008), 187–214.

[2] Bartnik, R., *The Mass of an Asymptotically Flat Manifold*, Comm. Pure Appl. Math. **39** (1986), 661–693.

[3] Bryant, R.L. and Salamon, S.M., *On the Construction of some Complete Metrics with Exceptional Holonomy*, Duke Math. J. **58** (1989), 829–850.

[4] Harvey, F.R. and Lawson, H.B., *Calibrated Geometries*, Acta. Math. **148** (1982), 47–157.

[5] Hitchin, N., *The Moduli Space of Special Lagrangian Submanifolds*, Ann. Scuola Norm. Sup. Pisa Cl. Sci. (4) 25 (1997), no. **3-4**, 503-515 (1998).

[6] Joyce, D.D., *Compact Manifolds with Special Holonomy*, OUP, Oxford, (2000).

[7] Joyce, D.D., and Salur, S., *Deformations of Asymptotically cylindrical Coassociative Submanifolds with Fixed Boundary*, Geometry and Topology, Vol. 9 (2005) Paper no. 25, 1115–1146.

[8] Kovalev, A.G., *Twisted connected sums and special Riemannian holonomy*, J. Reine Angew. Math. **565** (2003), 125–160.

[9] Leung, N. C., *Topological Quantum Field Theory for Calabi-Yau Threefolds and $G_2$-manifolds*, Adv. Theor. Math. Phys. **6** (2002), no. 3, 575–591.

[10] Lockhart, R.B., *Fredholm, Hodge and Liouville Theorems on noncompact manifolds*, Trans. A.M.S. **301** (1987), 1–35.

[11] Lockhart, R.B. and McOwen, R.C., *Elliptic differential operators on noncompact manifolds*, Ann. Scuola norm. sup. Pisa, Classe di scienze **12** (1987), 409–447.

[12] Melrose, R., *The Atiyah-Patodi-Singer Index Theorem*, A.K. Peters, Wellesley, MA, (1994).

[13] Melrose, R., *Spectral and Scattering Theory for the Laplacian on Asymptotically Euclidean Spaces*, in Ikawa M., editor, *Spectral and Scattering Theory*, Lecture notes in Pure and Applied Mathematics, vol. **161**, Marcel Dekker Inc., (1994).

[14] McLean, R.C., *Deformations of calibrated submanifolds*, Comm. Anal. Geom. **6** (1998), 705–747.

[15] Morrey, C.B. Jr., *Multiple Integrals in the Calculus of Variations*, Grund. der math. Wiss. **130**, Springer-Verlag, Berlin, (1966).

[16] Salur, S., *Asymptotically Cylindrical Ricci-Flat Manifolds*, Proceedings of American Mathematical Society, **134, No. 10** (2006), 3049–3056.

[17] Salur, S., *Deformations of Asymptotically Cylindrical Coassociative Submanifolds with Moving Boundary*, math.DG/0601420v2.

[18] Salur, S., *Deformations of Special Lagrangian Submanifolds*, Communications in Contemporary Mathematics **Vol.2, No.3** (2000), 365–372.

[19] Tian, G. and Yau, S.T., *Complete Kähler Manifolds with Zero Ricci Curvature I*, J. Amer. Math. Soc. **3** (1990), 579–609.

[20] Tian, G. and Yau, S.T., *Complete Kähler Manifolds with Zero Ricci Curvature II*, Invent. Math. **106** (1991), 27–60.



Department of Mathematics, University of Rochester, Rochester, NY, 14627
*E-mail address*: salur@math.rochester.edu

Department of Mathematics, University of Rochester, Rochester, NY, 14627
*E-mail address*: ajtodd@math.rochester.edu

Proceedings of $16^{th}$ Gökova
Geometry-Topology Conference
pp. 124 – 136

# Building symplectic manifolds using hyperbolic geometry

*Joel Fine and Dmitri Panov*

Abstract. This is a survey of the symplectic part of [7]. It is known that a hyperbolic manifold of even dimension is the base of a bundle whose total space admits a natural symplectic form ([11, 3]). We use this together with a construction resembling that of the Kummer surface to produce a simply-connected symplectic 6-manifold with vanishing first Chern class but no compatible complex structure. The role of the manifold-with-involution—the complex torus in Kummer's original construction—is in our case played by a beautiful hyperbolic 4-manifold discovered by Davis [4].

## 1. Introduction

This article gives a survey of some of the ideas of [7]. The goal is to describe the construction of a simply-connected symplectic 6-manifold which has vanishing first Chern class, but admits no compatible integrable complex structure. This is, to the best of our knowledge, the first such example to appear in the literature. The construction combines hyperbolic geometry and a resolution-of-singularities argument reminiscent of the Kummer surface. (The article [7] also gives an analogous construction of simply-connected complex threefolds with trivial canonical bundle, but which admit no symplectic forms. We do not describe this here.)

It follows from the theory of coadjoint orbits that every hyperbolic $2n$-manifold $M$ is the base of a fibration $X \to M$ whose total space is a symplectic manifold—see §2.1. This can also be seen as a special case of the work of Reznikov [11] concerning symplectic structures on twistor spaces. There, Reznikov considers more general metrics satisfying a curvature inequality. Since we need only the hyperbolic case, we do not describe this here. (For more information on this see also the discussion in [8], which focuses on this inequality in the 4-dimensional case and the related work of Davidov–Muškarov–Grantcharov [3].)

For $n \geq 3$, these symplectic manifolds have $c_1$ a positive multiple of the symplectic class—the symplectic analogues of Fano varieties, if you will. When $n = 2$, the first Chern class vanishes, giving symplectic "Calabi–Yaus".

The fundamental group of a symplectic manifold built this way is equal to the fundamental group of the hyperbolic base and so this does not give simply-connected compact examples directly. To get around this we use a construction analogous to that of the Kummer surface. We use a special hyperbolic 4-manifold, called the Davis manifold, to build a symplectic 6-manifold with an involution. Dividing by the involution kills the fundamental group and leaves a symplectic orbifold with vanishing first Chern class. We then resolve the singularities to give a smooth simply-connected symplectic manifold with $c_1 = 0$. This manifold has $b_3 = 0$ and so cannot support a Kähler structure with $c_1 = 0$. (Simply-connected Kähler manifolds with $c_1 = 0$ in fact have holomorphically trivial canonical bundle, hence $b_3 \geq 2$ because of the existence of a holomorphic volume form.)

## 2. From hyperbolic to symplectic manifolds

### 2.1. A hyperbolic coadjoint orbit

Given a Lie group $G$ and a coadjoint orbit $\mathcal{O} \subset \gamma^*$, it is a standard fact that $\mathcal{O}$ carries a natural symplectic structure. Given $u, v \in \gamma$ and $a \in \mathcal{O}$, the symplectic form at $a$ evaluated on the vectors in $T_a\mathcal{O}$ corresponding to $u, v$ is simply $a([u, v])$. We will apply this general fact to the group of isometries of hyperbolic $2n$-dimensional space.

The group of orientation-preserving isometries of hyperbolic space $H^m$ can be identified with $\mathrm{SO}(m, 1)$, the orientation-preserving linear isomorphisms of $\mathbb{R}^{m+1}$ which also preserve the Minkowski metric $x_0^2 - x_1^2 - \cdots - x_m^2$. The Lie algebra $\mathfrak{so}(m, 1)$ is the space of $(m + 1) \times (m + 1)$ matrices of the form

$$\begin{pmatrix} 0 & u^t \\ u & A \end{pmatrix} \tag{1}$$

where $u$ is a column vector in $\mathbb{R}^m$ and $A \in \mathfrak{so}(m)$. Those elements with $u = 0$ generate $\mathfrak{so}(m) \subset \mathfrak{so}(m, 1)$, which correspond to rotations of $H^m$ about a fixed point. The Killing form is definite on $\mathfrak{so}(m, 1)$ and so gives an equivariant isomorphism $\mathfrak{so}(m, 1) \cong \mathfrak{so}(m, 1)^*$. It follows that adjoint and coadjoint orbits can be identified.

We now restrict to the case when $m = 2n$ is *even*. Consider the adjoint orbit of

$$\xi = \begin{pmatrix} 0 & 0 \\ 0 & J_0 \end{pmatrix}$$

where $J_0 \in \mathfrak{so}(2n)$ is a choice of almost complex structure on $\mathbb{R}^{2n}$ (i.e., $J_0^2 = -1$). The stabiliser of $\xi$ is $\mathrm{U}(n)$. We write $Z$ for the corresponding orbit (or $Z_{2n}$ when we need to emphasise the value of $n$).

Thought of as a coadjoint orbit, $Z$ is symplectic. Since $Z_{2n} \cong \mathrm{SO}(2n, 1)/\mathrm{U}(n)$ we have $\dim Z_{2n} = n(n + 1)$. It fibres naturally over hyperbolic space $H^{2n} \cong \mathrm{SO}(2n, 1)/\mathrm{SO}(2n)$ with fibre isomorphic to the symmetric space $\mathrm{SO}(2n)/\mathrm{U}(n)$. This space can be identified with the collection of linear orthogonal complex structures on $\mathbb{R}^{2n}$ inducing a fixed orientation—that is to say orthogonal endomorphisms $J$ of $\mathbb{R}^{2n}$ with $J^2 = -1$ and such that a basis of the form $v_1, Jv_1, \ldots, v_n, Jv_n$ gives a fixed orientation. It follows that $Z$ is the twistor space of $H^{2n}$, although we don't directly use twistor methods here.

We next find an almost complex structure on $Z$ which is compatible with its symplectic structure. To do this we will use an equivariant description of the tangent bundle.

**Lemma 2.1.** *There is an isomorphism of* $\mathrm{U}(n)$*-representation spaces:*

$$\mathfrak{so}(2n,1) \cong \mathfrak{u}(n) \oplus \Lambda^2(\mathbb{C}^n)^* \oplus \mathbb{C}^n.$$

*Given a point* $z \in Z$ *with stabiliser* $\mathrm{U}(n) \subset \mathrm{SO}(2n,1)$ *there is a* $U(n)$*-equivariant isomorphism*

$$T_z Z \cong \Lambda^2(\mathbb{C}^n)^* \oplus \mathbb{C}^n, \tag{2}$$

*in which* $\Lambda^2(\mathbb{C}^n)^*$ *is tangent to the fibre of the projection* $Z \to H^{2n}$*.*

*Under this isomorphism, the symplectic form on* $T_z Z$ *is a positive multiple of the standard form on* $\Lambda^2(\mathbb{C}^n)^* \oplus \mathbb{C}^n$ *induced by the Euclidean structure on* $\mathbb{C}^n$*.*

*Proof.* There is a $\mathrm{U}(n)$-equivariant isomorphism $\mathfrak{so}(2n) \cong \mathfrak{u}(n) \oplus \Lambda^2(\mathbb{C}^n)^*$. To see this, write $\mathfrak{so}(2n) \cong \Lambda^2(\mathbb{R}^{2n})^*$. Given a choice of almost complex structure on $\mathbb{R}^{2n}$, any real 2-form $a$ can be written uniquely as $a = \alpha + \beta + \bar{\beta}$ where $\alpha \in \Lambda^{1,1}_{\mathbb{R}}$ is a real $(1,1)$-form and $\beta \in \Lambda^{2,0}$. Identifying $a$ with $(\alpha, \beta)$ gives a $\mathrm{U}(n)$-equivariant decomposition $\Lambda^2_{\mathbb{R}} \cong \Lambda^{1,1}_{\mathbb{R}} \oplus \Lambda^{2,0}$. Now, via the metric, $\Lambda^{1,1}_{\mathbb{R}}$ is identified with the skew-Hermitian matrices $\mathfrak{u}(n)$ and this gives the claimed isomorphism.

There is also an $\mathrm{SO}(2n)$-equivariant isomorphism $\mathfrak{so}(2n,1) \cong \mathfrak{so}(2n) \oplus \mathbb{R}^{2n}$. In the above form (1) the $\mathfrak{so}(2n)$ summand is given by $u = 0$ whilst the $\mathbb{R}^{2n}$ summand by $A = 0$. Combining these two completes the proof of the isomorphism (2).

Next we take care of the symplectic form. By $\mathrm{U}(n)$-equivariance, the form on $T_z Z$ must be proportional under (2) to the form induced by the Euclidean structure on $\mathbb{C}^n$. To show the constant of proportionality is positive, first check that the forms are genuinely equal in the case $n = 1$, where $\mathfrak{so}(2,1) \cong \mathfrak{u}(1) \oplus \mathbb{C}$ as a $\mathrm{U}(1)$-representation. This amounts to the fact that $Z_2 = H^2$ with symplectic form the hyperbolic area form. Next, use induction and the fact that the decompositions of $\mathfrak{so}(2n,1)$ and $\mathfrak{so}(2n+2,1)$ from above are compatible with the obvious inclusions of the summands induced by a choice of $\mathbb{C}^n \subset \mathbb{C}^{n+1}$. □

We define an $\mathrm{SO}(2n,1)$-invariant almost complex structure $J$ on $Z$ defined by declaring (2) to be a complex linear isomorphism. Notice that, by Lemma 2.1, $J$ is compatible with the symplectic structure on $Z$.

With respect to $J$ the tangent bundle splits $TZ = V \oplus H$ as a sum of complex vector bundles with $V \cong \Lambda^2 H^*$, corresponding to (2). The sub-bundle $V$ is those vectors tangent to the fibres of $Z \to H^{2n}$. We remark in passing that in the twistorial picture, this splitting is simply the decomposition of $TZ$ induced by the Levi–Civita connection of $H^{2n}$. From this point of view, the almost complex structure $J$ is the (non-integrable) "Eells–Salamon" structure [5], given by reversing the (integrable) "Atiyah–Hitchin–Singer" structure [1] in the vertical directions.

**Lemma 2.2.** $c_1(Z_{2n}) = (2-n)c_1(H)$.

*Proof.* This follows from $TZ = \Lambda^2 H^* \oplus H$ along with the fact that for any complex rank $n$ vector bundle $E$, $c_1(\Lambda^2 E) = (n-1)c_1(E)$. □

We next determine the symplectic class of $Z$. First, consider the restriction of the symplectic structure to the fibres of $Z \to H^{2n}$. It follows from (2) that the fibres are symplectic and almost-complex submanifolds. Moreover, the stabiliser $\mathrm{SO}(2n) \subset \mathrm{SO}(2n,1)$ of a point $x \in H^{2n}$ acts on the fibre $F_x$ over $x$ preserving both these structures. As mentioned above, $F_x \cong \mathrm{SO}(2n)/\mathrm{U}(n)$ is a symmetric space (the space of linear orthogonal complex structures on $T_x H^{2n}$ inducing a fixed orientation). The standard theory of symmetric spaces gives $F_x$ a symmetric Kähler structure. It follows from $\mathrm{SO}(2n)$-equivariance that this must agree with the restriction of the symplectic and almost complex structures from $Z$.

It is also standard that the symmetric Kähler structure on $\mathrm{SO}(2n)/\mathrm{U}(n)$ comes from a projective embedding. In this particular case, we can describe this as follows. Let $E \to F$ denote the "tautological" bundle over $F = \mathrm{SO}(2n)/\mathrm{U}(n)$. I.e., each point of $F$ is a complex structure on $\mathbb{R}^{2n}$; the fibre of $E$ at a point $j \in F$ is the complex vector space $(\mathbb{R}^{2n}, j)$. The bundle $\det E^*$ is ample and $c_1(\det E^*) = -c_1(E)$ is represented by the symmetric symplectic form on $F$.

In our situation, the splitting (2) tells us that the tautological bundle of the fibre $F_x$ is simply $H|_{F_x}$. It follows that on restriction to a fibre, the symplectic class agrees with $-c_1(H)$. However, topologically, $Z \cong F \times H^{2n}$ is homotopic to $F$. Since their fibrewise restrictions agree, it follows that $-c_1(H)$ is equal to the symplectic class of $Z$. Hence, writing $\omega$ for the symplectic form on $Z$, we have:

**Proposition 2.3.** $c_1(Z_{2n}) = (n-2)[\omega]$.

## 2.2. Compact quotients

The symplectic form and almost-complex structure on $Z$ as well as the splitting $TZ = V \oplus H$ and identification $V \cong \Lambda^2 H^*$ are $\mathrm{SO}(2n,1)$-invariant. It follows that all these arguments, in particular Proposition 2.3, apply equally to smooth quotients of $Z_{2n}$ by subgroups of $\mathrm{SO}(2n,1)$. Let $\Gamma \subset \mathrm{SO}(2n,1)$ be the fundamental group of a compact hyperbolic $2n$-manifold $M$. $\Gamma$ acts by symplectomorphisms on $Z_{2n}$ to give as quotient a symplectic manifold of dimension $n(n+1)$. It fibres $Z_{2n}/\Gamma \to M$ as the twistor space of $M$. In this way we can use compact hyperbolic manifolds to produce compact symplectic manifolds.

When $n = 1$, $Z_2/\Gamma = M$, the symplectic form is the hyperbolic area form and we have simply the recovered the hyperbolic surface itself.

When $n \geq 2$ the situation is more interesting. For $n = 2$, the fibre $\mathrm{SO}(4)/\mathrm{U}(2)$ is the 2-sphere. So the quotient $Z_4/\Gamma$ is a 6-manifold which is the total space of a 2-sphere bundle over $M$. Moreover, Proposition 2.3 says $c_1(Z_4/\Gamma) = 0$ and so hyperbolic 4-manifolds lead to symplectic "Calabi–Yau" manifolds. This was seemingly first observed in [8]. That article also describes a more general approach for constructing symplectic forms on the total space of $S^2$-bundles over 4-manifolds, which involves certain $\mathrm{SO}(3)$-connections, called *definite connections*. From that point of view, the Levi–Civita connection on $\Lambda^+ \to H^4$ is an example of a definite connection.

**Remark 2.1.** It will be important in what follows to note that when $n = 2$, not only do the symplectic manifolds have $c_1 = 0$, but they actually have a *preferred section of the almost canonical bundle.* This is because of the isomorphism (2) which in the case of $n = 2$ implies

$$\Lambda^3 T_z^* Z \cong \Lambda^2(\mathbb{C}^2)^* \otimes \Lambda^2(\mathbb{C}^2)$$

Thus the almost canonical bundle over $Z$ is not just trivial, but trivial in a natural $\mathrm{SO}(4,1)$-invariant way. It follows that the almost canonical bundle of the compact quotients also have natural trivialisations.

When $n \geq 3$, Proposition 2.3 says that compact hyperbolic $2n$-manifolds give compact symplectic manifolds for which $c_1$ is a positive multiple of $[\omega]$ (for $n \geq 3$). These might be considered as symplectic analogues of Fano varieties in algebraic geometry. Recall that $\dim Z_{2n} = n(n+1)$ so the lowest dimension of "symplectic Fano" which can be achieved in this way is 12. In this case, the fibration over $M^6$ has fibres $\mathrm{SO}(6)/\mathrm{U}(3) \cong \mathbb{CP}^3$.

We remark that, for $n \geq 2$, no compact manifold produced in this way can be Kähler. This is because, for $m > 2$ no discrete co-compact lattice in $\mathrm{SO}(m,1)$ can arise as the fundamental group of a compact Kähler manifold. In particular, the "Fanos" arising from hyperbolic manifolds of dimension at least 6 are non-Kähler. These examples (originally appearing in Reznikov's article [11]) are, to the best of our knowledge, the first non-Kähler symplectic "Fano" manifolds.

## 2.3. A Kähler description

In what follows, it will be important to have an alternative description of $Z$ in the case of $H^4$. In fact, $Z$ has an *integrable* complex structure which is compatible with the symplectic form. This complex structure is *not* $\mathrm{SO}(4,1)$-invariant (otherwise it would descend to all quotients) but is only invariant under the smaller group $\mathrm{SO}(4)$.

This can be described as follows. We denote by $Y$ the total space of $\mathcal{O}(-1) \oplus \mathcal{O}(-1)$ over $\mathbb{CP}^1$; write $\pi \colon Y \to \mathbb{CP}^1$ for the bundle projection. There is a map $p \colon Y \to \mathbb{C}^2 \oplus \mathbb{C}^2 \cong \mathbb{C}^4$ given by adding the maps $\mathcal{O}(-1) \to \mathbb{C}^2$. We can define a Kähler metric on $Y$ by pulling back the metrics from $\mathbb{CP}^1$ and $\mathbb{C}^4$:

$$\omega = \pi^* \omega_{\mathrm{FS}} + p^* \omega_{\mathbb{C}^4}.$$

The points $(x, y)$ and $(z, w)$ of $\mathbb{C}^2$ span the same line if and only if $xw - yz = 0$. It follows that the image of $p$ is the quadric cone $Q = \{xw - yz = 0\}$. The map $p \colon Y \to Q$ is called the small resolution of $Q$.

(Strictly speaking, there are two small resolutions, depending on whether one identifies $\mathbb{C}^2 \oplus \mathbb{C}^2 \cong \mathbb{C}^4$ by sending $((x, y), (z, w))$ to $(x, y, z, w)$ or to $(x, z, y, w)$. This latter gives another map $p' \colon Y \to Q$. There is no biholomorphism of $Y$ which swaps $p$ and $p'$, so they are genuinely inequivalent resolutions. For more details see, for example, the article [12] which gives a more thorough description of the small resolutions.)

The group $\mathrm{SO}(4, \mathbb{C})$ of complex linear isomorphisms of $\mathbb{C}^4$ with determinant 1 and preserving the quadratic form $xw - zy$ acts on $Q$. This action lifts to a holomorphic action on the resolution. If we want to also preserve the Kähler form, however, we must restrict to $\mathrm{SO}(4,\mathbb{C}) \cap \mathrm{U}(4)$. The Hermitian and complex forms determine a conjugation map on $\mathbb{C}^4$, the real subspace fixed by the conjugation is the copy of $\mathbb{R}^4 \subset \mathbb{C}^4$ where the two forms agree. From this it follows that $\mathrm{SO}(4,\mathbb{C}) \cap \mathrm{U}(4) \cong \mathrm{SO}(4)$. Here $\mathrm{SO}(4)$ acts on $\mathbb{C}^4 \cong \mathbb{R}^4 \oplus i\mathbb{R}^4$ by extending the standard action on $\mathbb{R}^4$ by complex linearity. (If we choose coordinates in which the complex quadratic form is given by $z_1^2 + \cdots + z_4^2$, the decomposition $\mathbb{C}^4 \cong \mathbb{R}^4 \oplus i\mathbb{R}^4$ just corresponds to $z = \Re z + i\Im z$.)

Projecting onto the real subspace gives a surjection $t\colon Q \to \mathbb{R}^4$. Away from the origin the fibres are two-spheres. The map lifts to the small resolution where it becomes a genuine $S^2$-bundle $t\colon Y \to \mathbb{R}^4$. This map is $\mathrm{SO}(4)$-equivariant and so we can see $Y$ in a similar fashion to $Z \to H^4$. Given this similarity the following is not surprising.

**Proposition 2.4.** *There is an* $\mathrm{SO}(4)$*-equivariant symplectomorphism between* $Y$ *and* $Z$.

This was proved first in [8] and then in a different way in [7]. We omit the details here.

## 3. A simply-connected symplectic Calabi–Yau

For the rest of this survey we focus on the construction of a simply-connected non-Kähler symplectic Calabi–Yau 6-manifold, which appears in [7]. The discussion above shows that a hyperbolic 4-manifold carries a 2-sphere bundle whose total space is naturally a symplectic "Calabi–Yau" manifold, with $c_1 = 0$. It is not possible to apply this directly to obtain a simply-connected compact example because, of course, compact hyperbolic manifolds have large fundamental group. To get around this problem we use a tactic resembling Kummer's construction of a K3 surface.

### 3.1. The Kummer construction

We begin by briefly recalling Kummer's construction. Let $A \cong \mathbb{C}^2/\mathbb{Z}^4$ be a complex torus. The involution $z \mapsto -z$ descends to $A$ where it fixes 16 points. The resulting quotient $A/\mathbb{Z}_2$ is singular, but simply-connected as a topological space (that is to say, we do not consider the orbifold fundamental group here). Moreover, the holomorphic volume form $\mathrm{d}z_1 \wedge \mathrm{d}z_2$ on $A$ is $\mathbb{Z}_2$-invariant and so $A/\mathbb{Z}_2$ has holomorphically trivial canonical bundle. To produce a smooth simply-connected surface with trivial canonical bundle, we now resolve the 16 singularities.

To give a local model for the resolution one can start with the model for the blow-up of a point $\mathcal{O}(-1) \to \mathbb{C}^2$. The involution $z \mapsto -z$ of $\mathbb{C}^2$ lifts to $\mathcal{O}(-1)$ where it now fixes the whole exceptional curve. The singularity of $\mathcal{O}(-1)/\mathbb{Z}_2$ is resolved by squaring the normal bundle to the exceptional curve. In other words, taking the "square-root" gives a map $\mathcal{O}(-2) \to \mathcal{O}(-1)/\mathbb{Z}_2$. Now composing with projection to $\mathbb{C}^2/\mathbb{Z}_2$ gives a resolution $\mathcal{O}(-2) \to \mathbb{C}^2/\mathbb{Z}_2$ in which the singularity has been replaced by a curve of self-intersection $-2$. Doing this locally at each of the 16 singular points of $A/\mathbb{Z}_2$ we get a smooth resolution $X \to A/\mathbb{Z}_2$. Moreover it is straightforward to verify that, since we have changed the topology only in codimension 2, the fundamental group is unaffected and so $\pi_1(X) = 1$.

An important feature of the model resolution $p\colon \mathcal{O}(-2) \to \mathbb{C}^2/\mathbb{Z}_2$ is that the pull-back $p^*(\mathrm{d}z_1 \wedge \mathrm{d}z_2)$, defined a priori away from the exceptional locus, actually extends to a global nowhere-vanishing holomorphic volume form on $\mathcal{O}(-2)$. This is described by saying the resolution is *crepant.* It follows that, when thought of as a form on $A/\mathbb{Z}_2$ (away from the singularities), $\mathrm{d}z_1 \wedge \mathrm{d}z_2$ pulls back to $X$ and extends to a holomorphic trivialisation of the canonical bundle. In other words, $X$ is a simply-connected complex surface with trivial canonical bundle.

## 3.2. The Davis manifold

In order to implement a similar construction in our situation, we first need a hyperbolic 4-manifold which can play the role of $A$. For this we use a beautiful manifold found by Davis [4]. (See also the article of Ratcliffe–Tschantz [10] where the homology of the Davis manifold is computed.)

The Davis manifold $M$ is built using a regular polytope called the 120-cell (or hecatonicosachoron). The 120-cell is a four-dimensional regular solid with 120 three-dimensional faces—the "cells"—each of which is a solid dodecahedron. Each edge is shared by 3 dodecahedra and each vertex by 4 dodecahedra. In total, the 120-cell has 600 vertices, 1200 edges and 720 pentagonal faces. Take a hyperbolic copy $P \subset H^4$ of the 120-cell in which the dihedral angles are $2\pi/5$. For each pair of opposite dodecahedral faces of $P$ there is a unique reflection in a hyperplane which identifies them. Gluing opposite faces via these reflections gives the Davis manifold, a hyperbolic 4-manifold $M$.

The central involution of $H^4$ which fixes the centre of $P$ preserves both $P$ and the identifications of opposite faces, hence it gives an isometric involution $\sigma$ of $M$. Our symplectic construction will begin with the resulting orbifold $M/\sigma$, which we call the *Davis orbifold.*

To analyse the fixed points of $\sigma$ it is helpful to use the so-called "inside-out" isometry of $M$ (defined in [10]). To describe this, note that $P$ can be divided up into 14400 hyperbolic Coxeter simplexes. The vertices of a simplex are given by taking first the centre of $P$, then the centre of one of its 120 3-faces $F$, then the centre of one of the 12 2-faces $f$ of $F$, then the centre of one of the 5 edges $e$ of $f$ and, finally, one of the two vertices of $e$. Denote by $v_1, v_2, v_3, v_4, v_5$ one such choice. The corresponding simplex has a isometry that exchanges $v_1$ (the centre of $P$) with $v_5$ (a vertex of $e$), $v_2$ (the centre of $F$) with $v_4$ (the centre of an edge of $F'$) and fixes $v_3$ (the centre of $f$). This isometry of the simplex extends to define the inside-out isometry of $M$, which commutes with $\sigma$.

**Lemma 3.1.** *The fixed set of* $\sigma$ *consists of* 122 *points. The quotient* $M/\sigma$ *is simply connected as a topological space.*

*Proof.* In the interior of the 120-cell there is only one fixed point, the centre, all other fixed points of $\sigma$ lie on the image in $M$ of the boundary of $P$. Let $F$ denote the image in $M$ of a three-dimensional face of $P$; $\sigma$ preserves $F$ and induces on it the symmetry of the dodecahedron given by inversion $x \mapsto -x$ with respect to its centre. So, once again, in the interior of $F$ there is only one fixed point, its centre. Considering all opposite pairs of three-dimensional faces of $M$ this gives 60 more fixed points of $\sigma$. All remaining fixed points are contained in the image in $M$ of the union of the 2-faces of $P$.

The symmetry $\sigma$ takes 2-faces to 2-faces. We claim next that $\sigma$ does not fix an interior point of any pentagonal 2-face. Assume for a moment that it does fix such a point. Then it would give an involution of the pentagon which would hence fix a vertex and so also the line joining the vertex to the centre of the polygon. The Davis manifold has two distinguished points, the centre and the image of all the vertices of the 120-cell. The assumption that $\sigma$ fixes an interior point of a pentagonal 2-face gives a $\sigma$-fixed tangent direction at the vertex point in $M$. However, the inside-out isometry exchanges the centre and vertex of $M$. Since $\sigma$ acts as $x \mapsto -x$ at the centre it does so also at the vertex and hence acts freely on the tangent space there. It follows that $\sigma$ does not fix an interior point of any 2-face.

The remaining fixed points are contained in the image in $M$ of the union of the edges of $P$. Under the inside-out involution of $M$, the middles of all edges are exchanged with centres of all 3-faces whilst the centre of $P$ is exchanged with the image in $M$ of the vertices of $P$. Since the inside-out isometry commutes with $\sigma$, this give an additional 61 fixed points of $\sigma$ making 122 in total.

We now turn to the (topological, not orbifold) fundamental group of $M/\sigma$. The map $\pi_1(M) \to \pi_1(M/\sigma)$ is surjective so we need to show its image is trivial. Consider the 60 closed geodesics $\gamma_i$ in $M$ going through the centre of $P$ and joining the centres of opposite faces. The deck transformations corresponding to these geodesics generate the whole of $\pi_1(M)$. Indeed, these deck-transformations take the fundamental domain $P$ to all its 120 neighbours. Now the result follows from the fact that every loop $\sigma(\gamma_i)$ is contractible. □

### 3.3. The model singularity and resolution

Locally, the singularities of $M/\sigma$ are modelled on the quotient of $H^4$ by $x \mapsto -x$. (Here $x$ is the coordinate provided by the Poincaré ball model of $H^4$.) This lifts to action on $Z \to H^4$ where the fixed locus is an $S^2$-fibre.

To understand the resulting singularity in the quotient of $Z$ we will use the Kähler description of Proposition 2.4. This says that we can identify $Z$ symplectically with $\mathcal{O}(-1) \oplus \mathcal{O}(-1)$. In this picture, the involution acts by $z \mapsto -z$ in the vector-bundle fibres of $\mathcal{O}(-1) \oplus \mathcal{O}(-1) \to \mathbb{CP}^1$. The quotient orbifold $\mathcal{O}(-1) \oplus \mathcal{O}(-1)/\mathbb{Z}_2$ is a Kähler Calabi–Yau orbifold with singular locus $\mathbb{CP}^1$ corresponding to the zero section. We next describe a crepant resolution of this singularity.

**Lemma 3.2.** *There is a crepant resolution*

$$\mathcal{O}(-2,-2) \to \mathcal{O}(-1) \oplus \mathcal{O}(-1)/\mathbb{Z}_2$$

*where $\mathcal{O}(-2,-2) \to \mathbb{CP}^1 \times \mathbb{CP}^1$ is the tensor product of the two line bundles given by pulling back $\mathcal{O}(-2) \to \mathbb{CP}^1$ from either factor.*

*Proof.* Blow up the zero section of $\mathcal{O}(-1) \oplus \mathcal{O}(-1)$ to obtain$\mathcal{O}(-1,-1) \to \mathbb{CP}^1 \times \mathbb{CP}^1$. The $\mathbb{Z}_2$-action lifts to this line bundle where it again has fixed locus the zero section and acts by $z \mapsto -z$ in the fibres. For such an involution on *any* line bundle $L$, the square gives a resolution $L^2 \to L/\mathbb{Z}_2$. Hence

$$\mathcal{O}(-2,-2) \to \mathcal{O}(-1,-1)/\mathbb{Z}_2 \to \mathcal{O}(-1) \oplus \mathcal{O}(-1)/\mathbb{Z}_2$$

gives the claimed resolution □

We should emphasise here that this is resolution is *holomorphic*. The total space of $\mathcal{O}(-2,-2)$ is a Kähler manifold with trivial canonical bundle and, away from the exceptional divisor, the map in Lemma 3.2 is a biholomorphism when we consider $\mathcal{O}(-1) \oplus \mathcal{O}(-1)/\mathbb{Z}_2$ with its *holomorphic* complex structure.

However, when constructing symplectic six-manifolds from hyperbolic four-manifolds, the relevant almost complex structure and volume form on $\mathcal{O}(-1) \oplus \mathcal{O}(-1)$ (and its $\mathbb{Z}_2$-quotient) are not the holomorphic ones; rather we use the $\mathrm{SO}(4,1)$-invariant almost complex structure and complex volume form from §2.1 and Remark 2.1. Lemma 3.2 can only be used to provide crepant resolutions of Calabi–Yau singularities modelled on the *holomorphic* geometry of $\mathcal{O}(-1) \oplus \mathcal{O}(-1)/\mathbb{Z}_2$ and *not* the $\mathrm{SO}(4,1)$-invariant almost complex structure and complex volume form of §2.1.

So, in order to apply Lemma 3.2 to resolve the singularities in a hyperbolic twistor space, we need to interpolate between the holomorphic structures near the zero section in $\mathcal{O}(-1) \oplus \mathcal{O}(-1)$ to the $\mathrm{SO}(4,1)$-invariant structures outside a small neighbourhood of the zero section. This interpolation is provided by the following result.

**Lemma 3.3.** *Let $Z_\delta$ denote the part of $Z \to H^4$ lying over a geodesic ball in $H^4$ of radius $\delta$. For any $\delta > 0$, there is an* $\mathrm{SO}(4)$*-invariant compatible almost complex structure $J$ on $Z$ and an* $\mathrm{SO}(4)$*-invariant nowhere-vanishing section $\Omega$ of the $J$-canonical-bundle such that:*

- *Over $Z_\delta$, $J$ and $\Omega$ agree with the standard holomorphic structures.*
- *Over $Z \setminus Z_{2\delta}$, $J$ and $\Omega$ agree with the* $\mathrm{SO}(4,1)$*-invariant structures from §2.1 and Remark 2.1.*

*Proof.* As is standard, an $\mathrm{SO}(4)$-invariant interpolation between the "inside" and "outside" Hermitian metrics gives the existence of $J$.

To produce $\Omega$ we start with a description of the $\mathrm{SO}(4)$-action away from the zero-section $Z_0$. The stabiliser of a point $p \in Z \setminus Z_0$ is a circle $S^1_p \subset \mathrm{SO}(4)$ and the orbit of $p$ is 5-dimensional (in fact, isomorphic as an $\mathrm{SO}(4)$-space to the unit tangent bundle of $S^3$). The lift of a geodesic ray out of the origin in $H^4$ meets each $\mathrm{SO}(4)$-orbit in a unique point, giving a section for the action. We interpolate between the holomorphic and hyperbolic complex volume forms along the relevant portion of this lifted ray and then use the $\mathrm{SO}(4)$-action to extend the resulting 3-form to the whole of $Z$. In order for this to work it is sufficient that at every point $p \in Z \setminus Z_0$ the action of $S^1_p$ on the fibre of the $J$-canonical-bundle at $p$ is trivial. But since the weight is integer valued and continuous it is constant on $Z \setminus Z_0$ so we can compute it for some $p$ outside of $Z_{2r}$ where everything agrees with the hyperbolic picture. Here we already have an $\mathrm{SO}(4)$-invariant (hence $S^1_p$-invariant) complex volume-form so the weight is zero as required. □

## 3.4. The twistor space of the Davis orbifold

With Lemmas 3.2 and 3.3 in hand, we can now take a crepant resolution of the twistor space of the Davis orbifold $M/\sigma$. Let $X \to M$ denote the twistor space of the Davis manifold The involution $\sigma$ lifts to an involution of $X$ which we still denote $\sigma$. $X/\sigma$ is a symplectic orbifold with singularities along 122 $\mathbb{CP}^1$s, each modelled on $\mathcal{O}(-1) \oplus \mathcal{O}(-1)/\mathbb{Z}_2$.

Let $\delta$ be a positive number small enough that the geodesic balls in $M$ of radius $2\delta$ centred on the $\sigma$-fixed points are embedded and disjoint. Then, by Lemma 3.3, on $X$ we can find a new almost complex structure $J$ and complex volume form $\Omega$ such that outside the geodesic $2\delta$-balls they agree with the hyperbolic structures coming from §2.1 and Remark 2.1, whilst inside the balls of radius $\delta$ they agree with the holomorphic structures coming from the holomorphic geometry of $\mathcal{O}(-1) \oplus \mathcal{O}(-1)$. It follows from the SO(4)-invariance in Lemma 3.3 that $J$ and $\Omega$ are $\sigma$-invariant.

In this way the quotient $X/\sigma$ is a symplectic orbifold with an almost complex structure and complex volume form which are modelled near the singular curves on the holomorphic geometry of $\mathcal{O}(-1) \oplus \mathcal{O}(-1)/\mathbb{Z}_2$. It follows from Lemma 3.2 that there is a resolution $\hat{X} \to X/\sigma$ in which the singular curves have been replaced by copies of $\mathbb{CP}^1 \times \mathbb{CP}^1$ with normal bundle $\mathcal{O}(-2,-2)$; moreover, $\hat{X}$ carries an almost complex structure $\hat{J}$ and complex volume form $\hat{\Omega}$ so that $c_1(\hat{X}, \hat{J}) = 0$.

Finally we need to define the symplectic structure on $\hat{X}$. Pulling back the symplectic form via $\hat{X} \to X$ gives a symplectic form on the complement of the exceptional divisors. To extend it we use a standard fact about resolutions in Kähler geometry. Given any neighbourhood $U$ of the zero locus in $\mathcal{O}(-2,-2)$, there is a Kähler metric on $\mathcal{O}(-2,-2)$ for which the projection to $\mathcal{O}(-1) \oplus \mathcal{O}(-1)/\mathbb{Z}_2$ is an isometry on the complement of $U$. (This amounts to the fact that the zero locus has negative normal bundle.)

So, in the model, the pull-back of the symplectic form extends over the exceptional divisor in a way compatible with holomorphic complex structure. Taking $U$ sufficiently small and doing this near all 122 exceptional divisors defines a symplectic form $\omega$ on $\hat{X}$ which is compatible with $\hat{J}$.

## 3.5. Non-Kählerity

It remains to prove that there is no compatible Kähler structure on $\hat{X}$; indeed there is no Kähler structure whatsoever with $c_1 = 0$. This will follow from the fact that $\pi_1(\hat{X})$ and $b_3(\hat{X})$ both vanish. To prove simple-connectivity we first recall a standard lemma.

**Lemma 3.4.** *Let $A$ and $B$ be two finite dimensional CW complexes and let $f\colon A \to B$ be a surjective map with connected fibres. Suppose that $B$ has an open cover by sets $U_i$ such that for any $y \in U_i$ the inclusion homomorphism $\pi_1(f^{-1}(y)) \to \pi_1(f^{-1}(U_i))$ is an isomorphism. Then the following sequence is right-exact:*

$$\pi_1(f^{-1}(y)) \to \pi_1(A) \to \pi_1(B) \to 0.$$

(This is a truncated version of the long exact sequence of homotopy groups associated to a Serre fibration. The proof is identical.)

**Lemma 3.5.** *$\hat{X}$ is simply connected.*

*Proof.* We first apply Lemma 3.4 to the map $X/\sigma \to M/\sigma$. The fibres are $S^2$s and we see that $\pi_1(X/\sigma) = 1$. Next we apply Lemma 3.4 to $\hat{X} \to X/\sigma$. This time the fibres are points or $S^2$s and we deduce that $\pi_1(\hat{X}) = 1$. □

To prove that $b_3(\hat{X}) = 0$ we invoke a lemma of McDuff on the cohomology of manifolds obtained by symplectic blow-ups.

**Lemma 3.6** (McDuff [9]). *Let $X$ be a symplectic manifold and $C \subset X$ a smooth symplectic submanifold of codimension $2k$. Let $\tilde{X}$ denote the blow-up of $X$ along $C$. Then the real cohomology of $\tilde{X}$ fits into a short exact sequence of graded vector spaces*

$$0 \to H^*(X) \to H^*(\tilde{X}) \to A^* \to 0$$

*where the first arrow is pull-back via $\tilde{X} \to X$ and where $A^*$ is free module over $H^*(C)$ with one generator in each dimension $2j$, $1 \le j \le k$.*

**Theorem 3.7.** $b_3(\hat{X}) = 0$.

*Proof.* Recall that $X \to M$ is the twistor space of the Davis manifold. We first blow up the 122 fibres which lie over the fixed points of $\sigma$ to obtain the new manifold $\tilde{X}$. It follows from Lemma 3.6 that pulling back cohomology via $\tilde{X} \to X$ induces an isomorphism $H^3(\tilde{X}) \cong H^3(X)$.

Next, notice that $\sigma$ lifts to $\tilde{X}$ and that $\hat{X} = \tilde{X}/\sigma$. We now show that $\sigma$ acts as $-1$ on $H^3(\tilde{X})$. To see this, consider the action of $\sigma$ on the Davis manifold $M$. It acts on $H^1(M)$ as $-1$ and hence also as $-1$ on $H^3(M)$. Now $X \to M$ is a sphere-bundle so, by Leray–Hirsch, $H^*(X)$ is a free module over $H^*(M)$ with a single generator in degree 2 corresponding to the first Chern class of the vertical tangent bundle. This generator is preserved by $\sigma$, so $\sigma$ acts as $-1$ on $H^3(X)$ and hence also as $-1$ on $H^3(\tilde{X})$. From this we deduce that $H^3(\hat{X}) = 0$. For if it contained a non-zero element, the pull-back to $\tilde{X}$ would be a $\sigma$-invariant element of $H^3(\tilde{X})$. □

**Theorem 3.8.** *A Kähler manifold with $b_1 = 0$ and $c_1 = 0$ has $b_2 \ge 2$.*

*Proof.* For a Kähler manifold, the vanishing of $b_1$ implies the Picard torus is trivial, i.e., a holomorphic line bundle which is topologically trivial bundle is necessarily holomorphically trivial. Now this combined with $c_1 = 0$ implies the canonical bundle is holomorphically trivial and so there is a holomorphic volume form. The real and imaginary parts of this form have independent cohomology classes and hence $b_3 \ge 2$. □

**Corollary 3.9.** *$\hat{X}$ admits no Kähler structure for which $c_1 = 0$.*

## 4. Generalising the construction

The last section of [7] describes a variety of questions which arise out of this construction. We restrict ourselves here to the following one: is it possible to use the same approach—hyperbolic orbifolds and resolution of singularities—to produce more examples of symplectic manifolds with vanishing first Chern class?

A collection of hyperbolic orbifolds is mentioned in [7] with relatively simple singularities (although not as simple as the isolated singularities of the Davis orbifold). In a forthcoming work we will show how the corresponding symplectic orbifolds can be resolved [6]. This produces a collection of simply-connected symplectic manifolds with $c_1 = 0$ and with arbitrarily large Betti numbers. It is as yet unknown whether a similar phenomenon can occur for Kähler Calabi–Yau manifolds.

If one were optimistic, however, one might hope that the potential for this construction is far greater. We need two things: the hyperbolic orbifolds and a way to resolve the resulting symplectic singularities. On the first topic, we mention a question due to Gromov: *Is every compact $n$-manifold homeomorphic to the quotient of $H^n$ by a discrete group of isometries?* (Of course, the group is allowed to have torsion.) In dimensions 2 and 3 this is already known to be true. In dimension 4 this problem is seemingly wide open.

On the second topic—crepant resolutions—we remark that for algebraic threefolds, crepant resolutions are always known to exist by a famous result of Bridgeland, King and Reid [2]. Even independently of the intended application described here, it would be interesting to know what holds in the symplectic setting.

If one were ambitious, one might hope to use this sort of construction to approach the question of what groups might appear as the fundamental group of a symplectic 6-manifold with $c_1 = 0$. Can any finitely presented group arise this way?

**Acknowledgements:** We are very grateful to the referee for pointing out an error in an earlier version of this article.

## References

[1] M. F. Atiyah, N. J. Hitchin, and I. M. Singer. Self-duality in four-dimensional Riemannian geometry. *Proc. Roy. Soc. London Ser. A*, 362(1711):425–461, 1978.

[2] T. Bridgeland, A. King, and M. Reid. The McKay correspondence as an equivalence of derived categories. *J. Amer. Math. Soc.*, 14(3):535–554 (electronic), 2001.

[3] J. Davidov, G. Grantcharov, and O. Muškarov. Almost complex structures on twistor spaces. In *Almost complex structures (Sofia, 1992)*, pages 113–149. World Sci. Publ., River Edge, NJ, 1994.

[4] M. W. Davis. A hyperbolic 4-manifold. *Proc. Amer. Math. Soc.*, 93(2):325–328, 1985.

[5] J. Eells and S. Salamon. Twistorial construction of harmonic maps of surfaces into four-manifolds. *Ann. Scuola Norm. Sup. Pisa Cl. Sci. (4)*, 12(4):589–640 (1986), 1985.

[6] J. Fine and D. Panov. Simply-connected symplectic 6-manifolds with vanishing first Chern class and arbitrarily large Betti numbers. To appear.

[7] J. Fine and D. Panov. Hyperbolic geometry and non-kahler manifolds with trivial canonical bundle. `arXiv:0905.3237`, 2009.

[8] J. Fine and D. Panov. Symplectic Calabi-Yau manifolds, minimal surfaces and the hyperbolic geometry of the conifold. *J. Differential Geom.*, 82(1):155–205, 2009.

[9] D. McDuff. Examples of simply-connected symplectic non-Kählerian manifolds. *J. Differential Geom.*, 20(1):267–277, 1984.
[10] J. G. Ratcliffe and S. T. Tschantz. On the Davis hyperbolic 4-manifold. *Topology Appl.*, 111(3):327–342, 2001.
[11] A. G. Reznikov. Symplectic twistor spaces. *Ann. Global Anal. Geom.*, 11(2):109–118, 1993.
[12] I. Smith, R. P. Thomas, and S.-T. Yau. Symplectic conifold transitions. *J. Differential Geom.*, 62(2):209–242, 2002.

Joel Fine, Départment de Mathématique, Université Libre de Bruxelles CP218, Boulevard du Triomphe, Bruxelles 1050, Belgique.
*E-mail address*: joel.fine@ulb.ac.be

Dmitri Panov, Department of Mathematics, Imperial College London, South Kensington Campus, London SW7 2AZ, United Kingdom.
*E-mail address*: d.panov@imperial.ac.uk

Proceedings of 16$^{th}$ Gökova
Geometry-Topology Conference
pp. 137 – 141

# Twisting 4-manifolds along $\mathbb{RP}^2$

*Selman Akbulut*

Abstract. We prove that the Dolgachev surface $E(1)_{2,3}$ (which is an exotic copy of the elliptic surface $E(1) = \mathbb{CP}^2 \# 9\bar{\mathbb{CP}}^2$) can be obtained from $E(1)$ by twisting along a simple "plug", in particular it can be obtained from $E(1)$ by twisting along $\mathbb{RP}^2$.

## 1. Introduction

Given a smooth 4-manifold $M^4$, what is the minimal genus $g$ of an imbedded surface $\Sigma_g \subset M^4$, such that twisting $M$ along $\Sigma$ produces an exotic copy of $M$? Here twisting means cutting out a tubular neighborhood of $\Sigma$ and regluing back by a nontrivial diffeomorphism. When $g > 1$ we don't get anything new (because by ([O], page 133)[1] any diffeomorphism of a circle bundle over $\Sigma_g$ can be isotoped to preserve the fiber, and hence it extends to the corresponding disk bundle). The case $g = 1$ is the well known "logarithmic transform" operation, which can change the smooth structure in some cases; in fact the first example of a closed exotic manifold found by Donaldson [D] was the Dolgachev surface $E(1)_{2,3}$ which is obtained from $E(1) = \mathbb{CP}^2 \# 9\bar{\mathbb{CP}}^2$ by two log transforms. The $g = 0$ case is not well understood, twisting along $S^2$ is usually called "Gluck construction" and we don't know if this operation changes the smooth structure of any orientable manifold, but there is an example of non-orientable manifold which the Gluck construction changes its smooth structure [A1]. The interesting case of $\Sigma = \mathbb{RP}^2$ was studied indirectly in [AY1] under the guise of *plugs*, which are more general objects. Recall that Figure 1 describes the tubular neighborhood $W$ of $\mathbb{RP}^2$ in $S^4$ as a disc bundle over $\mathbb{RP}^2$ (e.g. [A2]):

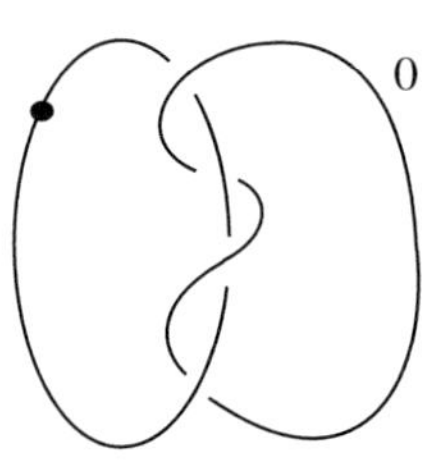


Figure 1. W

The author is partially supported by NSF grant DMS 0905917.
[1]We thank Cameron Gordon for pointing out this reference.

If we attach a 2-handle to $W$ as in Figure 2 we obtain an interesting manifold, which is the $W_{1,2}$ "plug" of [AY1]. Recall [AY1], a *plug* $(P, f)$ of $M^4$ is a codimension zero Stein submanifold $P \subset M$ with an involution $f : \partial P \to \partial P$, such that $f$ does not extend to a homemorphism inside; and the operation $N \cup_{id} P \mapsto N \cup_f P$ of removing $P$ from $M$ and regluing it to its complement $N$ by $f$, changes the smooth structure of $M$ (this operation is called a "*plug twisting*"). For example the involution $f : \partial W_{1,2} \to \partial W_{1,2}$ is induced from $180^0$ rotation of the Figure 2 , e.g. it maps the (red and blue) loops to each other $\alpha \leftrightarrow \beta$.

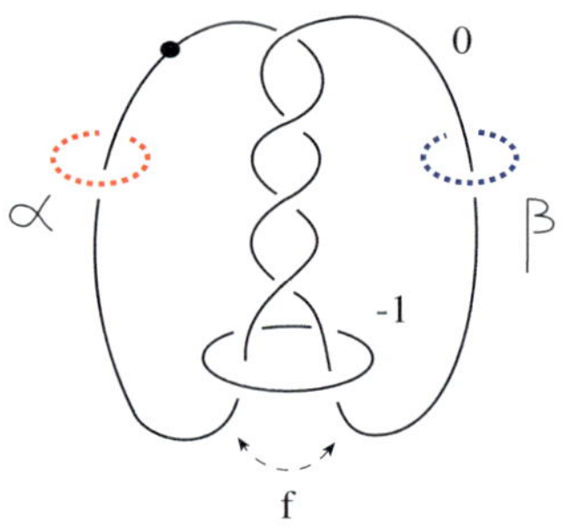


FIGURE 2. $W_{1,2}$

Notice that the twisting along $W_{1,2}$ is induced by twisting along $\mathbb{RP}^2$ inside (i.e. cutting out $W$ and regluing by the involution induced by the rotation). In [AY1] some examples of changing smooth structures via plug twisting were given, including twisting the $W_{1,2}$ plug. Here we prove that by twisting along a $W_{1,2}$ plug (in particular twisting along $\mathbb{RP}^2$) we can completely decompose the Dolgachev surface $E(1)_{2,3}$. The following theorem should be considered as a structure theorem for the Dolgachev surface complementing Theorem 1 of [A3], where it was shown that a "*cork twisting*" also completely decomposes $E(1)_{2,3}$.

**Theorem 1.1.** *$E(1)_{2,3}$ is obtained by plug twisting $E(1)$ along $W_{1,2}$, i.e. we have a decomposition $E(1) = N \cup_{id} W_{1,2}$, so that $E(1)_{2,3} = N \cup_f W_{1,2}$.*

*Proof.* By cancelling the 1- and 2-handle pair of Figure 2 we obtain Figure 3, which is an alternative picture of $W_{1,2}$. By inspecting the diffeomorphism Figure 2 $\mapsto$ Figure 3 we see that the involution $f$ twists the tubular neighborhood of $\alpha$ once, while mapping to $\beta$.

By attaching a chain of eight 2-handles to $-W_{1,2}$ (the mirror image of Figure 3) and a $+1$ framed 2-handle to $\alpha$, we obtain Figure 4, which is a handlebody of $E(1)$ given in [A3]. In Figure 4 performing $W_{1,2}$ plug twist to E(1) has the effect of replacing the $+1$-framed 2-handle attached to $\alpha$, with a zero framed 2-handle attached to $\beta$. Here the complement of $W_{1,2}$ in $E(1)$ is the submanifold $N$ consisting of the zero framed 2-handle (the cusp)

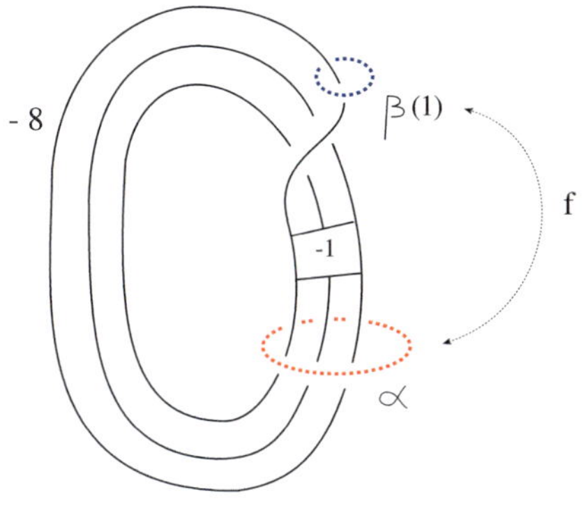


FIGURE 3. $W_{1,2}$

and the chain of eight 2-handles, and the plug twisting is the operation: $N \cup \alpha^{+1} \mapsto N \cup \beta^0$ (as seen from $N$).

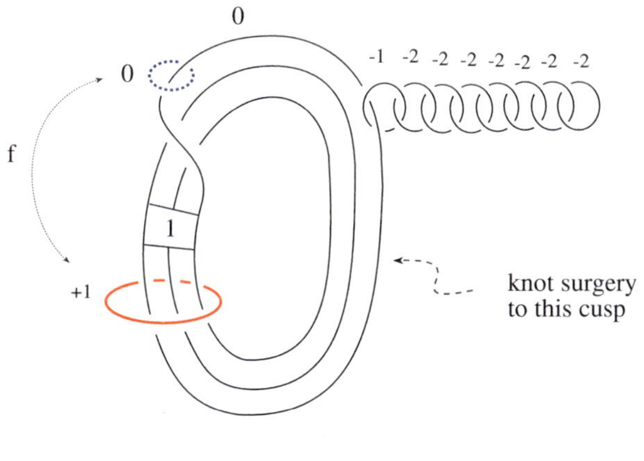


FIGURE 4. E(1)

Therefore the plug twisting of $E(1)$ along $W_{1,2}$ gives Figure 5. After sliding over $\beta$, the chain of eight 2-handles become free from the rest of the figure, giving a splitting: $Q\#8\bar{\mathbb{C}\mathbb{P}}^2$, where $Q$ is the cusp with the trivially linking zero framed cicle, hence we get $Q = S^2 \times S^2$. So the Figure 5 is just $S^2 \times S^2\#8\bar{\mathbb{C}\mathbb{P}}^2 = E(1)$.

Next notice that if we first perform a "knot surgery" operation $E(1) \mapsto E(1)_K$ by a knot $K$, along the cusp inside of Figure 4, and then do the plug twist along $W_{1,2}$ (notice

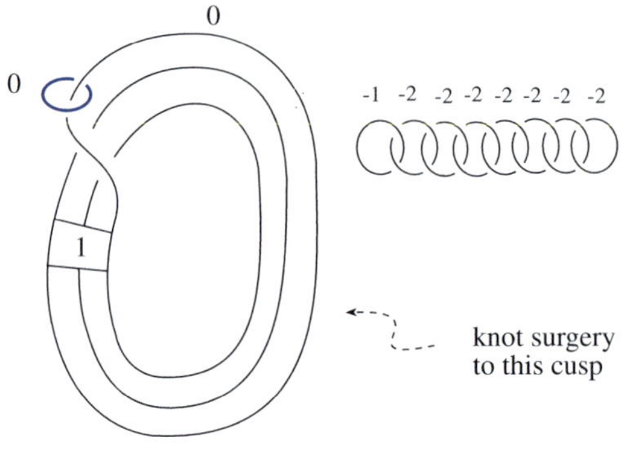


FIGURE 5

the cusp is disjoint from the plug since it lies in $N$) we get the similar splitting except this time resulting: $Q_K \# 8\bar{\mathbf{CP}}^2$, where $Q_K$ is the knot surgered $Q$. Notice the manifold $Q = S^2 \times S^2$ is obtained by doubling the cusp, and $Q_K$ is obtained by doing knot surgery to one of these cusps. In Theorem 4.1 of [A4] it was shown that when $K$ is the trefoil knot then $Q_K = S^2 \times S^2$. Also recall that when $K$ is the trefoil knot we have the identification with the Dolgachev surface $E(1)_K = E(1)_{2,3}$ (e.g. [A3]). □

**Remark 1.1.** If we could identify $Q_K$ with $S^2 \times S^2$ for infinitely many knots $K$ with distinct Alexander polynomials, we would have infinitely many transforms $E(1) \mapsto E(1)_K$ obtained by plug twistings along $W_{1,2}$. This would give infinitely many non-isotopic imbeddings $W_{1,2} \subset E(1)$, similar to the examples in [AY2]. In the absence of such identification we can only conclude that $W_{1,2}$ is a plug of infinitely many distinct exotic copies $E(1)_K$ of $E(1)$.

**Remark 1.2.** Recall that $\partial W$ is the quaternionic 3-manifold, which is the quotient of $S^3$ by the free action of the quaternionic group of order eight (e.g. [A2]):

$$G =< i, j, k \mid i^2 = j^2 = k^2 = -1, ij = k, jk = i, ki = j > .$$

This manifold is a positively curved space-form and an L-space (Heegaard Floer homology groups are trivial). Hence the change of smooth structure of $E(1)$ by twisting $W$ is due to the change of *Spin*$^c$ structures, rather than permuting the Floer homology by the involution as in [A3], [AD].

## References

[A1] S. Akbulut, *Constructing a fake 4-manifold by Gluck construction to a standard 4-manifold,* Topology, vol. 27, no. 2 (1988), 239–243.

[A2] S. Akbulut, *Cappell-Shaneson's 4-dimensional s-cobordism,* Geometry-Topology, vol.6, (2002), 425–494

[A3] S. Akbulut, *The Dolgachev surface,* arXiv:0805.1524v4 (2008).

[A4] S. Akbulut, *A fake cusp and a fishtail,* Turkish Jour. of Math 1 (1999), 19–31.

[A5] S. Akbulut, *Variations on Fintushel-Stern knot surgery,* Turkish Jour. of Math (2001), 81–92. arXiv:math.GT/0201156.

[AD] S. Akbulut, and S. Durusoy. *An involution acting non-trivially on Heegaard-Floer homology,* Fields Institute Communications, vol 47, (2005),1–9

[AY1] S. Akbulut and K.Yasui, *Corks, Plugs and exotic structures,* Jour of GGT 2 (2008), 40–82. arXiv:0806.3010

[AY2] S. Akbulut and K.Yasui, *Knotting Corks*, Journal of Topology (2009) 2(4), 823–839

[D] S.K. Donaldson, *Irrationality and the h-cobordism conjecture*, Journal of Differential Geometry 26 (1), (1987) 141–168.

[FS] R. Fintushel and R. Stern *Six lectures on 4-manifolds*, (2007), arXiv:math.GT/0610700v2.

[GS] R. Gompf and A. Stipsicz *4-manifolds and Kirby calculus*, (1999), AMS, GSM vol 20.

[O] P. Orlik *Seifert Manifolds*, LNM no.291 Springer-Verlag (1972)

Department of Mathematics, Michigan State University, MI, 48824
*E-mail address*: akbulut@math.msu.edu

16$^{th}$ GÖKOVA GEOMETRY-TOPOLOGY CONFERENCE
May 25 – 28, 2009

**List of Participants**

| | |
|---|---|
| Acu, Bahar | Middle East Technical University, Turkey |
| Akbulut, Selman | Michigan State University, USA |
| Argüz, N. Hülya | Middle East Technical University, Turkey |
| Arıkan, Fırat | University of Rochester, USA |
| Arslan, Aykut | Michigan State University, USA |
| Arslan, Havana | Ege University, Turkey |
| Atalan Ozan, Ferihe | Atılım University, Turkey |
| Auroux, Denis | Massachusetts Institute of Technology, USA |
| Baykur, İnanç | Brandeis University, USA |
| Bhupal, Mohan | Middle East Technical University, Turkey |
| Catanese, Fabrizio | Universität-Bayreuth, Germany |
| Ceyhan, Özgür | Max-Planck Institut, Germany |
| Chen, Weimin | University of Massachusetts-Amherst, USA |
| Coşkunüzer, Barış | Koç University, Turkey |
| Çelik, Fatih | Bilgi University, Turkey |
| Çiftçi, Ünver | Namık Kemal University, Turkey |
| Degtyarev, Alexander | Bilkent University, Turkey |
| Demir, A. Sait | Istanbul Technical University, Turkey |
| Dereli, Tekin | Koç University, Turkey |
| Diogo, Luis | Stanford University, USA |
| Dubrovin, Boris | SISSA, Italy |
| Efe, Barış | Michigan State University, USA |
| Erdoğan, Sultan | Bilkent University, Turkey |
| Etgü, Tolga | Koç University, Turkey |
| Erkuş, Soner | Dokuz Eylül University, Turkey |
| Finashin, Sergey | Middle East Technical University, Turkey |
| Ghiggini, Paolo | Université de Nantes, France |
| Gören, Yusuf | Bosphorus University, Turkey |
| Göktürk, Ali | Brown University, USA |
| Güğümcü, Neslihan | Middle East Technical University, Turkey |

| | |
|---|---|
| Güçlükan, Aslı | Bilkent University, Turkey |
| Gürbüzer, Kaan | Dokuz Eylül University, Turkey |
| İşim, Melike | Dokuz Eylül University, Turkey |
| Işık, Osman Raşit | Muğla University, Turkey |
| Itenberg, Ilia | IRMA, Université de Strasbourg, France |
| Juhász, András | University of Cambridge, UK |
| Kalafat, Mustafa | University of Wisconsin, USA |
| Karapazar, Şenay | Anadolu University, Turkey |
| Katzarkov, Ludmil | University of Vienna, Austria |
| Koçak, Şahin | Anadolu University, Turkey |
| Koca, Caner | SUNY Stony Brook, USA |
| Kouichi, Yasui | Kyoto University, Japan |
| Korkmaz, Mustafa | Middle East Technical University, Turkey |
| Matveyev, Rostislav | Universität Leipzig, Germany |
| McLean, Mark | ETH Zürich, Switzerland |
| Mikhalkin, Grigory | Université de Genève, Switzerland |
| Medetoğulları, Elif | Middle East Technical University, Turkey |
| Ozan, Yıldıray | Middle East Technical University, Turkey |
| Önder, Turgut | Middle East Technical University, Turkey |
| Özay, Mete | Middle East Technical University, Turkey |
| Özbağcı, Burak | Koç University, Turkey |
| Öztürk, Ali | İzzet Baysal University, Turkey |
| Pamuk, Mehmetcik | Koç University, Turkey |
| Pamuk, Semra | Koç University, Turkey |
| Panov, Dmitri | Imperial College London, UK |
| Petrunin, Anton | Münster University, Germany |
| Rossi, Paolo | SISSA, Italy |
| Saka, Işıl | Bosphorus University, Turkey |
| Sağlam, K. Nur | Middle East Technical University, Turkey |
| Salepci, Nermin | Koç University, Turkey |
| Salur, Sema | University of Rochester, USA |
| Santillan, Osvaldo | Trinity College, Ireland |
| Şengüler Çiftçi, Fatma | Groningen University, Holland |
| Stafa, Mentor | University of Rochester, USA |
| Taşkent, Selin | University of Rochester, USA |

| | |
|---|---|
| Tange, Motoo | Kyoto University, Japan |
| Tian, Gang | Princeton University, USA |
| Topkara, Mustafa | Middle East Technical University, Turkey |
| Tosun, Bülent | Middle East Technical University, Turkey |
| Uyanık, Çağlar | Middle East Technical University, Turkey |
| Ünal, İbrahim | University of Rochester, USA |
| Ünver, Emel | Ege University, Turkey |
| Yıldırım Yılmaz, Münevver | Fırat University, Turkey |
| Zabun, R. Arzu | Middle East Technical University, Turkey |
| Zhang, Ping | Eastern Mediterranian University, North Cyprus |

Made in United States
Troutdale, OR
08/17/2025

33725233R00093